NITRATES

NITRATES

The threat to food and water

NIGEL DUDLEY

First published in 1990 by
Green Print
an imprint of The Merlin Press Ltd
10 Malden Road, London NW5 3HR

ISBN 1 85425 012 4

1 2 3 4 5 6 7 8 9 10 : : 99 98 97 96 95 94 93 92 91 90 89

Phototypeset by Input Typesetting, London

Printed in England by Biddles Ltd., Guildford
on recycled paper

Contents

Preface

In April 1989, the EEC Commission gave the British government two months to produce detailed plans for meeting tough new community water regulations. Failure to do so would carry the risk of Britain being taken to the European Court. This marked the latest in a series of run-ins between Britain and the Community over water pollution issues. However, this time it gained greater significance, both because of the sudden upsurge in interest in green issues generally, and even more because of the intense debate surrounding water privatisation.

The single most important water pollution problem in Britain at present is nitrate contamination, arising from intensive forms of chemical farming. Nitrates have resulted in Britain being hauled before the European Court before. They have seriously dented public confidence in the purity of our drinking water supply. Questions about nitrate safety have involved a diverse and powerful collection of interest groups, ranging from the farming lobby through to anglers, naturalists, chemical companies, cancer specialists and organic farmers. These issues remain far from resolved.

Nitrates, or their derivatives, are important in other ways as well. Nitrate levels in fresh vegetables are causing growing alarm in some European countries. Nitrates in farm manure are an important source of air pollution, and thus of acid rain. And nitrates are increasingly suspected of being serious marine pollutants as well.

This book attempts to summarise what we know about the nitrate problem, and put it into the wider context of agricultural policy. It arises from a number of different sources. I first looked at nitrate pollution for the London Food Commission in 1985. Since then, I have continued to keep a watching brief on the debate for the Soil Association, and have been involved in the efforts of organic farmers to introduce their agricultural methods as an acceptable way of reducing nitrate leaching. Many people have contributed ideas and information over the past four years, and I am grateful to them all. I have benefitted particularly from conversations with David Baldock, Eric Brunner, Tim Lang, Julian Rose, Christopher Stopes and Lawrence Woodward. I also thank Jon Carpenter for steering this book through its various drafts, and for being unfailingly polite, even when I missed yet another deadline.

1: Spotlight on nitrates

Since the 1940s, changes in the philosophy and practice of agriculture have meant that there have been increasing quantities of nitrates in the environment. The increase in nitrates is inextricably linked with the intensification of farming. This includes inputs of nitrates from:

(1) liquid nitrate fertilisers;
(2) manures and slurries from intensive livestock units (factory farms);
(3) the effects of ploughing up pasture to create more arable land, and other modern cultivation methods.

As a result, nitrate levels have risen steeply in rivers and streams and underground aquifers. Concentrations have also increased in vegetable crops and livestock products.

This has a number of important side effects. An overdose of nitrates has helped impoverish freshwater life over large areas of Britain. There is growing evidence that nitrate pollution is damaging marine life as well. Derivatives of nitrates in livestock manure are implicated in a range of air pollution effects. And nitrates and their by-products expose us to both proven and suspected health risks.

Until recently the 'nitrates problem' has been

allowed to bubble along quietly without anything very much being done about it. The overwhelming priority given to maximising agricultural productivity allowed farmers and agrochemical companies to brush the issue to one side. Medical experts are still divided about the seriousness of nitrate's effects on our bodies, and this has allowed health issues to be ignored or patched over.

Thus, nitrate concentrations have been allowed to build up in food, water and the environment with very little public discussion. When action has been taken, it has usually involved the state-owned Water Authorities in extracting nitrates from water, rather than any attempt at tackling the problem at source. This has provided an important hidden subsidy to farmers who are causing high nitrate losses from their land. It breaks the 'Polluter Pays Principle' which is supposed to operate in Britain, whereby anyone causing pollution is legally obliged to pay for its cleaning up. The issue of increasing nitrate levels in food has been almost completely ignored in Britain, although it is viewed with concern in mainland Europe.

In the last five years, the problems associated with nitrate pollution have come to a head in Britain. There are a number of reasons for this:

(1) The scientific establishment has gradually (and in some cases reluctantly) admitted the strength of evidence linking intensive farming methods to the quantity of nitrates in the environment. Confirmation of this relationship has come from organisations as diverse as the Royal Society and the National Farmers Union, and from the government's own Department of the Environment.

(2) Several influential people and organisations have

drawn attention to the increasing levels of nitrates in freshwater. The Standing Technical Advisory Committee on Water Quality (STACWQ) has issued blunt warnings about the likelihood of further, dramatic increases unless radical changes in agriculture are introduced.

(3) Ecological damage in the North Sea is now believed to have been caused at least partly by agricultural pollution washing into coastal waters from rivers and estuaries.

(4) Other changes in the British and European political scene, including the debate about reducing crop surpluses in the European Community and the privatisation of water in Britain, have focused extra attention on the economic issues raised by nitrate pollution.

(5) Responsibility for resolving the problem of nitrates in water has now been taken out of the hands of the British government by the introduction of comprehensive European Community (EC) water quality legislation. The British government is already breaking EC laws, and even tighter restrictions on nitrate pollution are coming in the near future. About 1.3 million people in Britain sometimes receive tap water with nitrate levels higher than EEC safety regulations permit.

The political debate

Once the link between nitrate levels and farming methods was increasingly established and publicised, a number of political developments helped push nitrates further into the centre stage.

The first of these goes right back to the mid 1970s.

The 1975 EC Drinking Water Directive, which set maximum levels for nitrates in drinking water (amongst many other parameters) finally came into force in July 1985. Initially, the British government made what appeared to be a fairly clumsy attempt to avoid the EC legislation and was taken to the European Court as a result. British civil servants were then put under a great deal of pressure to find ways of reducing nitrate levels in water.

Two other political factors have been significant in the nitrates issue. Over the past few years, one of the most contentious aspects of European Community policy has undoubtedly been the EC's agricultural strategy, which has resulted in major food surpluses subsidised by the Common Agricultural Policy. These make up the infamous butter mountains and wine lakes.

To date, the only firm measure for tackling this problem is to pay farmers to take some areas out of production (so-called 'set aside'), or otherwise reduce levels of production ('extensification'). European taxpayers are faced with the Kafkaesque prospect of paying farmers to produce a surplus, paying them to leave some of their land idle to reduce this surplus, and paying for farming methods which result in a whole range of environmental pollutants, including nitrates. The incongruity is getting hard to explain away to the electorate. Now, there is fresh legislation, whereby 'Nitrate Sensitive Areas' (NSAs) are being designated for special attention and controls. Farmers in these areas face first voluntary and, probably, statutory control over the way in which they farm. NSAs will have a far more radical impact on agriculture than the current set-aside policy.

At the same time, the British government was in

the process of privatising the water supply. Until recently, as we have seen, the 'Polluter Pays Principle' has been ignored, and Water Authorities have paid the bill for reducing nitrates in water. Under private ownership (whatever its other disadvantages) there will be far less complacency with respect to this further, hidden, subsidy of British agriculture. The behind-the-scenes arguments between people controlling water and those controlling agriculture will inevitably spill out into open conflict about liability. The decision about who pays to remove the growing levels of nitrate contamination from drinking water is becoming a hot political issue.

Faced with scientific disquiet, pressure from the European Community and the need to do some window dressing as it puts up a pollution-ridden water industry for sale, the British government has been forced to rethink its agricultural policy. Nitrate pollution is central to these issues.

Other nitrate issues

Although nitrate levels in water are receiving the lion's share of attention at the moment, nitrates in food are also coming under increasing scrutiny. This includes nitrates deliberately added to meat products, and nitrates which build up in vegetables as a side-effect of the way in which they are grown.

Developments in some other European countries, where nitrate levels in vegetables are controlled by law, are increasing the pressure on the British government to take similar action. But, as yet, there has been little consumer interest in the issue of nitrates in vegetables.

In addition, nitrates are now known to be contributing to other environmental problems as well. In later chapters we look at the impact of nitrates from fertilisers and, especially, animal manures on local air pollution, and acid rain. And we examine disturbing new evidence that nitrates are now also damaging some of the marine life around our coasts.

2: Nitrogen, nitrates and the cycle of life

Nitrates aren't bad in themselves. Problems arise when too much nitrate gets into the wrong places. This chapter looks at where nitrogen and nitrates occur in nature, their particular relationship to agriculture, and why we are increasingly losing nitrates from farmland into the wider environment.

Nitrogen is essential for life. It is a key constituent of *amino acids*, which themselves act as building blocks for *proteins* in both plants and animals. Nitrogen is often the rate-limiting factor controlling plant growth in Britain. This means that, within limits, the more nitrogen that is added to the soil, the greater the crop yield. The rate at which nitrate fertiliser is applied thus often controls the rate of crop growth. This relationship has been responsible for many of the current problems connected with nitrates and farming.

Nitrate is a negatively charged ion which is readily soluble in water. In nature, nitrates are usually formed by the nitrification of nitrous oxides, themselves formed from ammonium, which in turn is formed from nitrogen. Some of the main chemical interactions are summarised in Figure 1.

Nitrate moves through the ecosystem in a number of

ways, known collectively as the nitrogen cycle. These include:

- Biological nitrogen fixation of nitrogen in the air by prokaryotic bacteria. These bacteria live in the roots of leguminous plants, such as clover and peas, and allow the plants to trap nitrogen from the atmosphere.
- Assimilation of fixed nitrogen. Plants pick up nitrogen from the soil via their roots, and animals obtain nitrogen by eating plants.
- Release of ammonia from plant and animal organic nitrogen through the action of bacteria (this is known as ammonification).
- Nitrification of ammonia, first to nitrite and then to nitrate.
- Reduction of nitrate to nitrogen oxides, and nitrogen to ammonium, by bacteria.

Although these reactions might seem complicated at first reading, the basic steps involved are quite simple. Plants collect nitrate from the soil, and some species also collect nitrogen from the atmosphere with the help of bacteria. Animals get nitrogen by eating plants, or by eating other animals which have themselves eaten plants. And various bacteria convert nitrogen into a number of different compounds to allow the cycle to continue.

Human influence on nitrate levels

Ever since the beginning of settled agriculture, maintenance of soil fertility has been a key element of farming practice. Nitrates are often the rate-limiting factor in British soils. Until the latter part of this century, nitrate levels in Britain have been main-

Figure 1

THE CHEMISTRY OF NITRATES

The Nitrogen Cycle has the following main components:

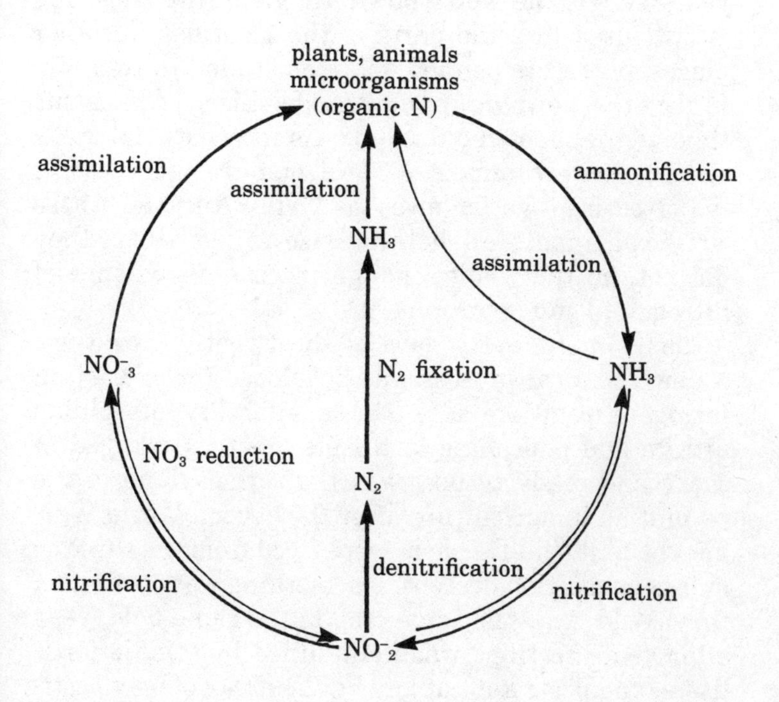

A generalised nitrogen cycle, showing the major processes involved, from *The Nitrogen Cycle in the United Kingdom*, The Royal Society, 1983.

tained by regular spreading of manure, known as 'dunging', and by leaving land fallow as part of a regular rotation. Clover and other legumes on the fallow land 'fix' nitrogen from the air.

There were regional variations to this basic framework of soil care. In Western Scotland, for example, large amounts of kelp, a brown seaweed, were collected after winter storms and heaped up on raised vegetable plots known as 'lazy beds', to provide fertility for the year. This is the well known 'tangle of the isles' and is still used in some parts of the Hebrides. In other places, pigs were penned on areas of land in rotation, so that their own dung provided fertiliser at the same time as they rooted out weeds. As international trade became more common, seabird manure (guano) was imported from as far away as Chile. Although these 'artificial' inputs all helped raise soil fertility, they did not, on the whole, change the overall balance of nitrogen in the environment.

Then, in the early days of the twentieth century, a new chemical process was developed for extracting nitrogen from the air. The availability of soluble nitrate and phosphate fertilisers, neatly packaged in sacks and ready to use, was the corner stone of the revolution in agriculture after the Second World War. For the first time, farmers were freed from the three or four year rotation of crops. If something was profitable, they could grow the same crop in the same field, year after year. Fertility was maintained by soluble fertilisers, and pests kept at bay by the array of new pesticides available.

The incoming Labour government of 1945 introduced a vast programme of intensification of British agriculture, which they regarded as an industry in urgent need of modernisation. Farmers were encouraged to maximise productivity. Farmers discovered that, within reason, the more fertiliser they added, the greater the yield of crops they produced. Fertilisers were so cheap that it was always worth adding enough for the best growing season, so that surplus applications became the norm.

Some years later, when Britain joined the European Community, a further spur to intensification was added by the Common Agricultural Policy, which guaranteed prices for the main crops. No longer constrained by either the natural carrying capacity of the land, or the risk of glutting the market place, farmers used all the technology available to produce as much food as they could from the land.

However, this new impetus meant that, more often than not, there was far more fertiliser added than was actually required by the crops. The fate of this surplus, and the effects of other aspects of agricultural intensification, make up the bulk of this book.

How nitrates enter and leave an area of land

The most important aspects of the nitrogen cycle from our point of view are the ways in which nitrates enter or leave an area of land, and the ways in which we affect this, especially by our methods of farming. Although there are several ways in which nitrate can enter or leach from soil naturally, the largest losses in agricultural areas of industrialised countries are now caused by our farming methods.

Inputs: Natural ways for nitrogen to enter an area of land include biological nitrogen fixation, as described above, and lightning discharges. Lightning was once believed to be a major source of nitrates but is now thought to be fairly negligible in countries with a climate like that in Britain.

Human, or anthropogenic, sources of nitrates include artificial nitrogen fertilisers, which are made from nitrogen extracted from the air in chemical factories, sewage discharged to land or freshwater and industrial and vehicular pollution, including deposition of both gases and 'acid rain'. Nitrates are also recycled within farmland, for example in manure from livestock. Because the animals are almost certainly fed on bought in food, this also equals a net input.

Outputs: Outputs of nitrogen include denitrification, which is the biological conversion of nitrates to gases such as nitrogen, and leaching of nitrates through the soil into surface waters and underground aquifers. Both these have important ecological effects, and we will come back to them later.

The relative importance of different sources of nitrates

In agricultural areas, by far the largest sources of nitrates are artificial fertilisers which make up almost 80 per cent of the total. In 1987, annual agricultural use of nitrates in Britain included: nitrate fertilisers at 1.6 million tonnes a year; manure at 0.4 million tonnes; and sewage sludge at 0.02 million tonnes. The proportions are 79 per cent from fertilisers, 20 per cent from manure and 1 per cent from sewage sludge. Other sources include rain, silage effluent, straw feed waste, seeds and biological nitrogen fixation.

An estimate of the percentage of each is given in Table 1. These figures were first worked out in 1978, and are thus rather out of date, but still appear to be the best available, and were used in the House of Lords

Select Committee report in July 1989. Since then, the proportion of nitrate fertilisers has increased sharply as compared to livestock manure.

Table 1: Relative importance of different nitrate sources in Britain

Input source	*Percentage of total UK input*
Artificial fertiliser	43
Livestock manure	38
Rain	10
Biological N fixation	6
Sewage	1
Seeds	1
Straw	0.5
Silage effluent	0.5

Source: Royal Society (1983); *The Nitrogen Cycle in the United Kingdom.*

The only other sizeable input is from rain, and this usually includes a large proportion of nitrogen oxide pollution from power stations, vehicles etc, and also from farming itself. The air pollution effects of farming are looked at in Chapter 5.

Losses of nitrates from farmland

The role of modern chemical farming is crucial to the nitrate story. Whilst just over half the nitrate leaves farmland incorporated in crops and grass, a great deal is also lost through leaching and volatilisation (i.e. rapid evaporation), as Table 2 shows. (Again, these figures are now rather outdated.)

Table 2: Output of nitrates from British farmland

Output source	*Percentage of total UK output*
Crops and grass	51
Volatilisation (1)	22
Denitrification, immobilisation etc	15
Leaching	12

Note: (1) volatilisation is rapid evaporation. It includes livestock manure (90 per cent); crop wastes (8 per cent); and sewage (2 per cent).
Source: Royal Society (1983); The Nitrogen Cycle in the United Kingdom

The role of nitrate fertilisers

Use of nitrate fertilisers is critical to the nitrates problem. The importance of nitrate fertilisers increased dramatically after 1940, as farming became more intensive. Use of nitrogen on farms tripled during the Second World War. Fertiliser application trebled again between 1960 and 1980, and use continued to increase 4–5 per cent per year throughout much of the 1980s.

Today, consumption of nitrate fertilisers runs at about 1.5 million tonnes a year. Farmers spend about £600 million on liquid nitrate fertilisers each year. About half the fertiliser is applied to grassland, 40 per cent to wheat and barley, and the rest to other crops, especially oilseed rape, potatoes and sugar. In England and Wales, average application rates are 180 kg/hectare for wheat, 122 kg/hectare for barley and 105 kg/hectare for grass.

With the Common Agricultural Policy (CAP)

guaranteeing fixed prices for cereals, it has made economic sense for farmers to add enough nitrate to maximise output in the best possible growing conditions. This means that, almost invariably, far more nitrate is added than is actually used in plant growth, leading to losses through run-off and leaching.

One of the most significant changes in recent years is that grazing pasture has also been treated with increasing doses of nitrate fertilisers. These are added to maximise growth of grass and thus further boost milk yields. Unfortunately this also increases leaching (and increases the amount of surplus milk which we pay for and have to throw away). Some analysts suggest that yields could be increased by applications of up to 300 kg/hectare, which is about three times what is usually being applied today.

Livestock manure

The pollutant effects of livestock manure are more complicated. Manure voided directly in fields can be a useful fertiliser, but tends to suffer heavy losses through volatilisation. Manure from factory farm units or barns is often produced in quantities too large for it all to be used as fertiliser near the source, but it is not valuable enough to be worth transporting. In these cases, manure is either dumped onto the land in excess – which causes extra leaching – or stored in slurry tanks. Intensive livestock farming has increased pollution from livestock manure by increasing the density of farm animals in some areas. (Manure also has an effect on air pollution, as described in Chapter 5.)

Ploughing and land management

In addition to nitrate added in fertilisers and manure, nitrate which has previously been 'locked into' the soil can be released by changing management practices. Examples of ways to increase nitrate losses include ploughing up permanent pasture and drainage. The importance of these 'management losses' is sometimes under-estimated. However, they have also recently become the focus for a lot of attention as the fertiliser lobby have tried to blame virtually all nitrate increases on management rather than fertiliser use.

Modern farming methods have vastly increased the amount of nitrogen entering agricultural land. Much of this increase is used up in the rapid growth of crops and livestock, but a sizeable proportion is lost from the system. The places this nitrate ends up, and the effects that this 'extra' nitrate has on human health and the environment, are still the subject of intense debate. In the next chapter, we try to untangle the threads relating to how nitrate from various aspects of modern farming makes its way into our food and drinking water.

3: Nitrates in our diet, and how they get there

Increasing levels of nitrates in the environment mean that we eat and drink greater quantities than our ancestors did. The major sources for most of us are vegetables, meat, dairy products and water. In this chapter, we look at each in turn.

Nitrates in vegetables

Most people think of nitrate as mainly a contaminant of water. In fact, most British people eat considerably more nitrates in food than they drink in water. Vegetables usually provide the largest single source, and so we start our survey of dietary nitrates by looking at how and why nitrate levels have increased in some types of vegetables over the last few years.

Plants take up inorganic nitrates through their roots. Plants of the pea family (legumes), including beans, clovers and vetches, also have the ability to extract or 'fix' atmospheric nitrogen through the action of symbiotic bacteria in their root nodules. Nitrate which they collect but do not use immediately for the synthesis of proteins, amino acids or nucleic acids, is stored within plant cells. Although nitrates build up in natural crops, concentrations appear to be increased by heavy use of artificial fertilisers. In these conditions, plants may be unable to carry out photosynthesis fast enough to get rid of all the surplus nitrates. They can, therefore, arrive on the dinner table still containing high nitrate levels.

Nitrate levels in vegetables build up even more when crops are harvested while they are still growing quickly. At this time, they still often have large nitrate reserves, especially when they are heavily treated with fertiliser. Growing crops in greenhouses during the winter, which has become very common because consumers have been told to expect to eat the same food all year round, can result in abnormally high nitrate concentrations. Research in Switzerland has found that nitrate concentrations can be up to twelve times higher than normal in winter-grown lettuce, for example. Even vegetables grown organically, without the use of soluble nitrate fertilisers, build up high nitrate levels if grown under glass in the winter.

Different parts of the plant store different amounts of nitrate. As a rule, stem and leaf tissues tend to accumulate the most nitrates, followed by roots and lastly flowers and fruit. Some crop species are more prone to building up nitrates than others, and environmental conditions can also affect nitrate levels.

Nitrate concentrations vary between crops

Nitrate levels can vary enormously both between species and within any given type of vegetable. For example, when the Royal Commission on Environmental Pollution investigated the nitrate concentrations in different British vegetables and root crops, they found levels in *spinach* varying from 310 mg/kg to 3,809 mg/kg. Research in Switzerland has found roughly the same levels in *winter lettuce*, and even higher concentrations in some of the less well known crops; *corn salad* reaching 5,250 mg/kg and *mangold* having a maximum of 5,200 mg/kg.

Britain's own Department of the Environment reports that nitrate levels in some commercially grown vegetable crops can reach as high as 10g/kg fresh weight; about 8 per cent of dry matter.

Amongst those species which can build up especially high nitrate levels are *spinach*, *lettuce* and *beetroot*, along with leafy vegetables like *cauliflower* and *cabbage*. On the other hand, some species apparently never build up high nitrate levels even in conditions of a large excess; these include *legumes* like broad beans and runner beans, *carrots*, *sugar beet*, *onions* and *leeks*. Although *potatoes* usually have relatively low nitrate concentrations, we eat so many potatoes, chips and crisps in Britain that they make up 15 per cent of the nitrate content in food (i.e. not including water) in our average diet.

Nitrates also vary with strains of crop, and growing methods

In the last few years, research projects at the Swiss National Research Institute for Biological Husbandry, and at Elm Farm Research Centre in Berkshire, have suggested that modern cultivars accumulate higher nitrate levels than traditional varieties in some crops. Modern crop varieties are bred for maximum yield, and tend to build up large reserves of nitrate within their cells, which can still be present when they are harvested.

Research in Switzerland has also compared nitrate levels built up in different cultivars of lettuce, under conditions of different fertiliser application. Concentrations ranged from 47 mg/kg for Neckarriese lettuce grown with 100 kg of compost per hectare to 1,666 mg/kg for Benita lettuce grown with 240 kg NPK of fertiliser per hectare; about 35 times greater.

Variation also occurs within particular crops and growing systems. The results given above are only an indication of the range of nitrate levels likely to be found. It would be misleading to put too much emphasis on published figures of nitrate concentrations in different crop species. Nonetheless, some indication of the crops and vegetables with the highest nitrate accumulation can be inferred from research over the past few years.

Some of the results have been collected in Table 3.

In a number of other countries, including Austria, Holland and Switzerland, Maximum Admissible Concentrations (MACs) have already been set for nitrate levels in some vegetables. West Germany has official 'guide values'. Britain has none of these. It is known that vegetables sold in Britain sometimes contain nitrates exceeding these maximum levels. But the best that the government working party on the issue could come up with by way of reducing nitrate intake from vegetables was to suggest that people eat less vegetables!

Nitrates in meat

Meat products account for a fairly small proportion of dietary nitrate, supplying an average of about 9 per

Table 3: Nitrate Concentrations in a range of fresh vegetables (mg/kg)

Vegetable	*Range*	*Mean Value*
Lettuce (summer)	160–3,100	1,100
Lettuce (winter)	2,000–10,200	
Cabbage (summer)	0–1,520	
Cabbage (winter)	40–1,480	
Cauliflower	40 ->254	
Brussel sprouts	0 – 170	
Celery leaves	2,665–5,580	
Spinach	35–6,700	
Mangold	2,500–5,200	
Fennel	1,000–1,650	1,400
Leek	300–4,480	
Carrot	0–2,850	
Potato	0 – 143	
Beetroot	630–6,800	
Turnip	45–2,900	420
Green bean	45 – 840	
Pea	0 – 15	5
Onion	0–2,250	125
Cucumber	70 – 140	
Kohlrabi	100–1,200	
Radish	925–3,850	
Corn salad	50–5250	

Notes: Some of these figures are amalgamated from several different experiments

Sources: Royal Commission on Environmental Pollution, 7th Report *Agriculture and Pollution*, (1979), HMSO, London; Vogtman, H. and Biedermann, R., *The Nitrate Story: No end in sight*, (1985) Elm Farm Research Centre, Newbury, Berkshire; Greenwood, D. J. and Hunt J., 'Effect of nitrogen fertiliser on the nitrogen content of field vegetables grown in Britain', *Journal of Science of Food and Agriculture*, vol. 37; London Food Commission, *Food Adulteration*, (1988), Unwin Hyman (contains other references).

cent of food intake according to the Royal Commission on Environmental Pollution. However, the contribution is unusual in that nitrate is added deliberately. (Nitrates are also added to a few cheeses.) It is also significant that there is now a school of thought which says nitrates in meat are a more serious health problem than nitrates from other sources. These questions are discussed in more detail in Chapter 4.

Nitrates and nitrites have been used as curing salts and preservatives in food, and especially in meats, for hundreds of years. The official reason for using these additives is that they prevent the growth of the bacterium *Clostridium botulinum* which causes botulism, a potentially lethal form of food poisoning. However, a bonus from the food industry's point of view is that nitrates give an additional red colour to mammal meat.

Legislation controlling the amounts of nitrites in meat has existed since 1940, and similar regulations covering nitrates were introduced in 1962.

The current UK regulations date from 1982. They define maximum levels of added nitrates and nitrites for *cured meats, sausages, bacon, ham, tongue, tinned* and *pressed meat, turkey loaves, frozen pizza* and certain *cheeses* such as *Edam, Gouda, Havarti, Danablu, Saint-Paulin* and *Tilsit.*

The maximums are interesting when compared with levels found in vegetables. The maximum level for nitrate (E251: sodium salt, E252: potassium salt) is 500 mg/kg and for nitrite (E249: potassium salt, E250: sodium salt) is 200 mg/kg. Thus twenty times as much nitrate as the maximum permitted in meat has been found in unusually highly contaminated lettuce. Nitrates and nitrites are completely banned as deliberate additives in infant food.

There have been several official recommendations that nitrates and nitrites should be eliminated altogether as additives in meat and dairy products. In 1978, the Ministry of Agriculture's Food Additives and Contaminants Committee (now replaced by the Food Advisory Committee) stated that:

> . . . we reiterate our earlier recommendation that every effort should continue to be made to eliminate the use of nitrate and to reduce nitrite levels as soon as practicable.

Despite such an unequivocal recommendation from MAFF over a decade ago, very little concrete evidence of this concern can be seen. There is no public information about whether the 1982 residual nitrate and nitrite limits are being adhered to or not. The fact that we are breaking EEC limits on nitrates in water is not reassuring.

Neither does any progress appear to have been made towards the supposed goal of phasing out nitrates altogether, as demanded by MAFF in 1978. A literature search, carried out for the London Food Commission in 1986, found no papers at all published in Britain about developing alternatives to nitrate for botulism control. The dozen or so research papers that were unearthed came from the USA, Japan and West Germany, where nitrate levels in meat are regarded with much greater concern. Indeed, the National Academy of Sciences, based in Washington DC, published a whole report on the subject in 1981.

In 1986, MAFF's Nitrate Coordination Group stated that they will research alternatives to nitrates for use as additives. These include ascorbates, which are derived from vitamin C.

Nitrates in water

Nitrate which has run off fields, or been leached through the soil, is likely to end up in freshwater, and hence get into drinking water. Nitrate contamination occurs in both surface waters, such as rivers and lakes, and in underground aquifers, or groundwater. In areas where nitrate contamination is particularly bad, nitrates from water overtake nitrates from vegetables as the main source in people's diets. Most of the excess nitrates in water come from chemical agriculture, although sewage discharges are also important.

The government's own Nitrate Co-ordination Group, discussed in more detail below, admitted the clear link between nitrates and agriculture in its report, released in 1987:

> A relationship between intensive arable cropping sustained by increasingly large nitrogen fertiliser applications and high rates of nitrate leaching to groundwater was inferred. High nitrate concentrations were accompanied by elevated levels of other solutes, notably sulphate, chloride and some trace elements, derived either directly from fertilisers or mobilised from soils by farming practices.

However, what is much more contentious is exactly where in the farming cycle the nitrates found in surface and ground waters actually come from. This might seem fairly esoteric. In practice, it is. But arguing about whether nitrates come principally from ploughing up fields, livestock density, or fertiliser applications has given the agrochemical companies valuable breathing space while the government has carried out more research, rather than acting to reduce nitrate contamination. Identification of precise causes of nitrate pollution are, therefore, extremely important.

Nitrates in surface waters

The occurrence of high, and increasing, levels of nitrates in surface waters has been recognised since the 1950s. The extent to which nitrate pollution has already affected freshwaters in Britain is discussed in Chapter 5. Nitrates in surface waters come from five main sources:

(1) discharge of treated sewage effluent;
(2) ploughing up grassland, which releases nitrates previously held in vegetation or in the root layer;
(3) improved land drainage, which can result in the conversion of organic nitrogen to inorganic nitrogen;
(4) release of silage effluents or slurry from livestock units;
(5) use of artificial (mostly liquid) nitrate fertilisers.

The relative importance of each has been the subject of intense debate, mainly because it has a direct bearing on who pays to clean up nitrates in water.

The farming establishment, and the agrochemical industry, insist that the releases are predominantly due to ploughing, or other cultivation techniques, rather than to the amount of fertiliser applied. Ploughing up land during and after the Second World War increased arable acreage by 8 million acres. Continual cultivation, and the regular ploughing and reseeding of pasture, have also increased the amount of land being turned over. Each time this happens, nitrates are lost from the system, and this is undoubtedly an important factor in general nitrate levels in water.

However, other researchers have argued that a proportion of the fertiliser itself is being leached downwards. And no one disagrees that the increase in nitrate leaching results from the intensification of agriculture over the past forty years. Some of these sources are examined in more detail in the following pages.

The extent of nitrate pollution in British freshwaters

The Standing Technical Advisory Committee on Water Quality (STACWQ) was set up to monitor water quality in Britain. Its members were never a group to make inflammatory statements without being fairly sure of their ground. Therefore, they really put the cat among the pigeons in 1985, when their annual report contained predictions about nitrate levels in water which made most of the environmentalists' warnings looks positively optimistic.

Most substantial British rivers have their nitrate levels regularly measured by a series of Harmonised Monitoring Stations. By assembling time series data for 25 British rivers, the Committee looked at what had happened to nitrate levels over the last few years, and made predictions for the future. The evidence they collected showed a 'clear and consistent' picture of rising nitrate levels in freshwater.

Concentrations had, on average, doubled over the past twenty years. In 1983, when they carried out the survey, they found an average increase of 0.65 mg nitrate/litre of water/year. They concluded that:

> without a substantial change in existing trends, further increases in average nitrate concentration of 8–13 mg/l may be expected over the next 10–20 years.

The STACWQ report caused a storm of controversy. The National Farmers Union, spurred on by their richer and more combatitive members, argued that the rate of increase in nitrate levels was tailing off and that, therefore, the STACWQ figures were too pessimistic. The Soil Association, Friends of the Earth, and the London Food Commission all became involved in the nitrates issue, and produced reports about various aspects of water pollution.

The government, already under pressure from the European Community, set up a Nitrate Coordination Group specifically to assess the impact of nitrates in water, and ways of reducing nitrate pollution from agriculture. It was chaired by Michael Healey of the Department of the Environment (DoE), and included fifteen other members from the Civil Service, official or semi-official bodies like the British Geological Survey and the Freshwater Biological Association, and commercial interest groups including the National Farmers Union and the Fertiliser Manufacturers' Association. There was no representation from nature conservation groups or specifically from health experts.

The NCG had problems in meeting their objectives and, probably, in reaching agreement about what their report should conclude. The Group eventually reported back a year late, in a DOE Pollution Paper (number 26) dated 1986 but not actually released until January 1987. This report is a key document in the history and politics of nitrate pollution and we will be returning to it several times in the course of the book. Although the NCG attempted a comprehensive overview of the problems posed by nitrates in water, and made recommendations about action which could be taken to alleviate nitrate pollution, political events in Britain and Europe meant that it was out of date almost as soon as it appeared.

The Nitrate Coordination Group report included an update of the STACWQ time series data. The official government paper provided slightly more optimistic figures. The DoE data indicated that the upward trend was not universal, at least in the early 1980s. Fifteen of the sampling points actually contained lower average nitrate concentrations than they did in the late 1970s. However, the remaining ten, concentrated especially in East Anglia, had recent nitrate levels higher than ever before.

There have been a number of factors, including perhaps a slowing down of the cultivation of new land, which could have helped stabilise nitrate levels in areas where farming is not quite so intensive. However, in the 'grain belt' of eastern England, nitrate pollution is continuing to climb steeply. In addition, this data only looks at surface waters, and so doesn't take account of problems which may be caused by the gradual leaching of water into underground aquifers.

Nitrates in groundwater

Groundwater makes up about 30 per cent of public drinking water supply in Britain, although in some areas the proportion is much higher. Groundwaters have traditionally been particularly free of contamination with nitrates.

The contamination of underground aquifers is the nitrate pollution issue which has most caught the imagination of politicians and media. This is largely because of the 'time bomb' theory, which suggests that nitrates which have already been released, and are leaching slowly downwards through the soil and rock, will lead to a steady but inexorable rise in the nitrate concentrations of some groundwaters well into the twenty-first century.

A proportion of the nitrate lost from arable fields or pasture leaches downwards, out of reach of plant roots and through successive soil and rock layers. Eventually, most of this reaches underground aquifers. There is a very large potential for groundwater contamination in many parts of Britain, and increased nitrate levels in groundwater have already been found in many areas.

The 'residence time' of water in many British underground aquifers is very long – often more than thirty years. Water extracted for drinking today is quite likely to come from sources established before the main onset of intensive chemical farming. In these waters, many of the potential effects of nitrate pollution will literally not have arrived yet.

In addition, some researchers believe that it takes nitrates many years to leach down to groundwater reservoirs; this is the essence of the time bomb theory. If the theory is true, the current increases in nitrate levels in water owe more to post-war ploughing than to the use of fertilisers. The effects of fertilisers could, therefore, be seen several decades into the future, leading to even greater increases.

How fast does nitrate leach into groundwaters?

The time bomb theory has caused a storm of controversy. There is some evidence that it could be occurring in areas where the rock is of a type which discourages fast leaching. In arable areas with a permeable catchment, the unsaturated (i.e. higher) zone frequently has nitrate concentrations of over 100 mg/litre. If this pollution continues to move downwards, then a steep rise in nitrates can be expected in the saturated (lower) zone as well.

However, there is disagreement about the speed with which water moves downwards. Water can move rapidly through cracks and fissures or slowly through pores in the rock layer.

Studies in Jurassic limestone found that there were fairly short, seasonal, fluctuations in groundwater nitrate concentrations. This suggests that nitrate can move fairly quickly through the fissures.

On the other hand, research in chalk indicated that 85 per cent of the water moved through pores in the rock at a speed of 0.8–1.1 metres a year, thus taking 20–30 years to reach groundwater. Some measurements in Triassic sandstone suggest downwards movement of about 2 metres a year.

We still don't know if the time bomb theory is correct or not. Initial studies using tritium as an indicator suggested that nitrate might be leaching downwards very slowly. However, some researchers think that tritium does not percolate at the same rate as nitrates, thus making any comparison meaningless. It has also been suggested that nitrate might be broken down and rendered harmless by biological processes on the way down anyway, although this is speculative.

To some extent, the time bomb debate is irrelevant to the central question. We already know that nitrates are polluting groundwater to a worrying extent. As the DoE's Nitrate Coordination Group concluded at the end of 1986:

> The available evidence . . . suggests that there will be a continuing slow rise in groundwater nitrate

> concentrations in most unconfined aquifers. . . in regions of lowest rainfall (parts of eastern and central England) many groundwater nitrate concentrations are likely in the long term to exceed 100 mg NO3/l. In other parts of Britain, with the exception of the highest rainfall areas of the west, a large number of groundwater sources in unconfined aquifers are currently estimated as likely to reach equilibrium concentrations in the range 50–100 mg NO3/l.

A recent DoE study, based on data from 75 sites, showed that around two thirds had increased nitrate levels, with the rest apparently static, but with substantial fluctuations. The worst affected supplies appear to be in the Severn-Trent, Southern, Anglian and Yorkshire Authorities.

Thus the government's own survey work suggests that almost all unconfined aquifers are likely to exceed the EC's limits on nitrate concentration. Huge areas will have twice as much, or more, of the permitted limit. A paper published by the Anglian Water Authority in 1984, *Nitrate in Drinking Water: Implications for the UK Water Industry*, is even less compromising:

> With nitrate peaks deriving from first time ploughing of arable land and addition of nitrate fertiliser still not having reached the water table in many areas, the future trends seem certain to show a continued increase in groundwater nitrogen concentrations. With current levels of nitrate fertiliser usage and 40–50 per cent leaching of nitrate from the soil, groundwater nitrate concentrations may ultimately reach some 35–40 mg/litre (N) in all outcrop aquifers overlain by arable land.

Point source pollution from slurry or silage effluent

Another increasingly important source of nitrate pollution comes from the relatively large amounts of slurry released from storage tanks, and from leaks of silage liquor, used as winter feed.

Slurry from intensive livestock units (i.e. factory farms) is often stored in tanks until it can be safely disposed of on farmland. This can involve fairly large quantities. An adult dairy cow can be responsible for sixteen and a half cubic metres of cow pat every year, containing, amongst other things, over 50 kg of nitrates. Concentrated animal wastes in slurry are 80–100 times more polluting than human sewage.

Silage is even more polluting, averaging 200 times more pollution than an equivalent amount of human waste.

Slurry causes many livestock farmers a headache and, in recent years, there has been a growing number of incidents where slurry tank walls are breached and the liquid escapes. This happens either because the tank has been badly made or because it is deliberately opened to get rid of excess slurry. Fines for illegal dumping are so low that they are unlikely to put off anyone determined to get rid of their slurry problem illegally.

Although slurry pollution is fairly unimportant in terms of overall pollution levels, it can cause intense local pollution, leading to fish kills. For example, the River Torridge, in Devon, was once a famous salmon river, and the setting for Tarka the Otter. The otters have long disappeared, as have many of the fish, and South West Water wrote in 1986 that

> in the last ten years large fish mortalities have been caused by discharges from agricultural sources and traders,

and described the potential load of silage effluent used to feed Devon's average dairy herd (58 cattle) as

> equivalent to that of a community of 10,800 inhabitants.

Indeed, the situation is now so serious that some Water Authorities are identifying it as a major cause of river deterioration in some areas. In a 1985 survey, *River Quality in England and Wales*, almost all the regional Water Authorities reported a steep increase in farm pollution incidents, including a ten-fold increase in just five years in the North West region. The Severn Trent Authority complained that:

> of particular concern is the impact of repeated pollution incidents in rural areas. Silage liquors and animal wastes are so polluting that even small quantities can be sufficient to affect water quality. Repeated minor entries of such wastes are having a chronic effect upon river quality. . . In a few cases, Class 1B streams (good quality) have been reduced to Class 3 (poor quality) because of repeated pollution incidents.

The Severn Trent Authority identified these incidents as 'the predominant cause of fish mortality', as did the Wessex Water Authority. The North West Water Authority found that over half the length of deteriorating river stretches in their area were on account of agricultural drainage 'from both diffuse and discrete sources'. And the South West Water Authority points to a 'continuing increase in reported farm pollution incidents' as the most obvious indication of changing farm practice.

This trend has continued. Slurry pollution incidents from farms have increased nine times in the last ten years, and reached their highest ever levels in 1988. The worst offenders were intensive pig and cattle rearing units. Recorded pollution incidents from farms reached 4,141 in 1988, almost three times the number of incidents reported in 1979.

Even so, the recorded incidents understate the extent of the problem. Surveys in the Severn Trent and South West areas showed that 46 per cent or more of dairy farms have pollution problems; 'dispelling the view that only a small number of farms were responsible for reduced river quality' according to Severn Trent Water.

There remains little to deter the unscrupulous or careless farmer. Despite the vast increase in the number of incidents, prosecutions actually fell in 1988, from 225 the previous year to 148. This means that farmers causing pollution from slurry only have a 1 in 28 chance of being prosecuted. In some areas the risk of prosecution is even lower. For example, the South West Water Authority only prosecuted 6 per cent of offending farmers. Reported average fines ranged from £165 to £630. When a farmer in Welshpool allowed silage to leak into a drinking water reservoir, it cost the water authority £4,000 to repair the damage, but the polluter was only fined £300, less than a tenth of even the clear-up costs.

However, this lax situation may soon change, under regulations laid down through the new Water Act. These enforce minimum construction standards for new silage and slurry stores, and the standards will be extendible to existing stores on a case by case basis. More importantly, an old rule whereby pollution resulting from 'good agricultural practice' is deemed under law not to be an offence, will be scrapped.

Nitrates from sewage pollution

It shouldn't be assumed that all the blame for nitrate pollution of our rivers falls on the farming community. In 1986, a major pollution issue surfaced when it was revealed that 23 per cent of Britain's sewage works, over a thousand separate operations in all, were breaching consent with regard to the amount of pollutants they released. This inevitably includes many nitrates from sewage, adding to the total load in Britain's rivers. In practice, it was revealed that the government had turned a blind eye to this situation until water privatisation, and increasing pressure from the European Community, forced ministers into action.

In November 1988, the Department of the Environment invited Water Authorities to apply for exemption from the rulings, either as an interim measure or permanently in the case of some small sewage works. The move was highlighted in the media, and caused a storm of controversy. By 1988, the total number of sewage works in breach of consent had fallen to 17 per cent of the total, some 751 in all, but without any chance of bringing most others into line before privatisation. Lord Crickhowell, chair of the new National Rivers Authority advisory committee, accused the authorities and the DoE of forming 'cosy incestuous relationships that have inhibited effective action' to clean up sewage discharges.

One of the most important factors currently in dispute is a plan for new upper tier limits for key pollutant indicators, which are never to be exceeded, to avoid one-off pollution incidents. An important upper tier parameter is biological oxygen demand which is controlled mainly by the amount of nitrate and phosphate in the water. At present, the DoE is still prevaricating over the issue, and the Water Authorities are pushing hard for a much lower limit than is currently being proposed.

Which source is most important?

Over the past few years, the agrochemical industry has been jumping through hoops in their attempts to show that fertilisers are not responsible for the nitrates leaching into freshwaters. Their arguments are increasingly being discounted. Although ploughing, and other forms of land management, are undoubtedly important in increasing leaching, they are all fuelled by the overall intensification of agriculture in the past four decades. And this intensification is, in turn, fuelled by heavy use of fertilisers. The Agricultural and Food Research Council pointed out this link in evidence to a House of Lords Select Committee in 1989 when they wrote that:

> The nitrate that leaks from arable land comes mainly from the breakdown of soil organic matter and crop residues. *As more fertiliser is added to produce more crop, more organic matter and organic residues are available for breakdown.* [My emphasis].

So how much nitrate is there likely to be in your water supply?

In 1987, the Department of the Environment estimated that 1.6 million people drank water which

Table 4: Areas affected by high nitrate levels in the water

Water Authority	*Supply Area*
Anglian WA	Barrow
	Dersingham and Snettisham
	Docking
	Heacham and Hunstanton
	Moulton and Kennett
	Wisbech
	Habrough
	Aswardby – Saltersford
	Binbrook
	Bully Hill – Barnoldby
	Bully Hill – Otby
	Clay Hill – Drove Lane
	Sleaford – Drove Lane
	Potterhanworth – Waneham Bridge
	Saltersford
	Waneham Bridge – Saltersford
	Ely and Littleport
	Habrough, Covenham
	Bowthorpe
	Heigham – Thorpe
	Otby
	Little London
	Winterton Holmes – Barrow
	Branston Booths
	Waddington – Glentham – Ulceby
	Isleham
Severn-Trent Water	Parts of Leamington, Kenilworth, Warwick and Stratford
	Parts of Stourbridge, Dudley
	Parts of Workshop, Mansfield Woodhouse (Bassetlaw and Mansfield DC)
	Malvern Hills and Wyre Forest DC
	Parts of Wrekin DC (Lilleshall source)
	Parts of South Shropshire DC
	Parts of Wrekin DC (Puleston Bridge)
	Parts of Newark DC
South Staffordshire Waterworks Co	Parts of Lichfield DC, Sutton Coldfield, North Warwickshire CC and Walsall Metropolitan areas
South West Water Authority	St Mary's Island
Lee Valley Water Co.	Part of Luton

Source: Nitrate in Water, Select Committee on the European Communities, 18 July 1989.

sometimes exceeded EEC nitrate limits. This water came from 28 surface water sources and 142 ground water sources, almost entirely from England. Andrew Lees, of Friends of the Earth, puts the figure even higher, at nearer 4 million.

This figure is increasing. In evidence to the House of Lords Select Committee report on *Nitrate in Water*, published in July 1989, the Water Authorities Association stated that rising nitrate levels would affect increasing proportions of their supplies. In the Severn Trent Water Authority, for example, about 7 per cent of total supplies now exceed 50 mg/litre, but this proportion is expected to double over the next decade. In East Anglia, the proportion is currently 4 per cent, but is expected to rise to 15 per cent by 1995.

Table 4 shows which areas are likely to be worst affected, according to the government's own figures. In the next chapter, we look at some of the things which consuming large quantities of nitrates can, or may, do to our health.

Nitrates in Europe are high as well

Rising nitrate levels are not confined to the UK. Throughout Europe and North America, intensive agriculture continues to cause nitrate leaching and water pollution.

For example, in *France*, over a million people are already drinking water with nitrate levels between 40 and 50 mg/litre and levels are rising fast. The latest estimates suggest that up to 10 million people could be drinking water exceeding Maximum Admissible Concentrations by 1990 or 1995. In *West Germany*, nitrate levels already exceed 50 mg/litre in many regions, and continue to increase fast. In *Italy*, levels of up to 100 mg/litre have been discovered, although the total number of people affected is still thought to be relatively low (about 100,000). In the *Netherlands*, almost a quarter of groundwater sources are expected to experience problems, and private wells have been measured with nitrate concentrations reaching a staggering 400 mg/litre.

The continent-wide phenomenon has alerted the European Community to the need for radical changes. Their action, in turn, is now putting pressure on the British government.

4: What nitrates do to us

On 27th November 1985, Donald Acheson, the Chief Scientist at the Department of Health and Social Security (now the Department of Health), wrote to all Medical Officers, GPs and teaching hospitals about possible health effects of nitrate in drinking water. He was trying to dispel persistent rumours that high nitrate levels can cause cancer, and to explain the UK health legislation governing nitrate levels in water, which mainly rests on avoiding risk to young children. Amongst other things, his detailed, two page letter said that:

> Although a theoretical risk of a relationship between nitrate and cancer remains, the epidemiological evidence, looked at as a whole, gives no support to the suggestion that nitrate is a cause of cancer of the stomach, or of any other organ, in the United Kingdom. . .

The fact that a letter was written at all shows the depth of concern with which the British government viewed the nitrates issue by the mid 1980s. Since then, the debate has become even more intense, and nitrate pollution maintains a high position on the political agenda. Unfortunately, questions about the health effects of nitrate pollution are shrouded in disagreements, poor research data and rhetoric. Separating out the facts is difficult; making a firm decision is probably still impossible. This chapter summarises the various issues as they stand at the moment, and makes a tentative assessment.

Nitrates, nitrites and nitrosamines

Any health problems connected with high nitrate levels are unlikely to be caused directly by the nitrates themselves, but by two of their breakdown products; nitrites and nitrosamines. Except in the case of a very serious overdose, nitrates are fairly non-toxic, and are rapidly excreted through the kidneys.

Nitrite is the reduced form of nitrate. The conversion is undertaken by bacteria known as microbial nitrate-reductases. It can occur both inside the stomach and mouth, and in food itself, through poor storage or the re-heating of some pre-packed foods. Most nitrite is probably built up by bacteria living inside the mouth.

Once nitrites reach the stomach, they can be converted to nitrosamines. This involves a reaction between secondary or tertiary amines or amides and nitrites, in the presence of catalysts such as thiocyanate or halogenoids. (Thiocyanate levels will be about four times higher than normal in the stomach of regular tobacco smokers, meaning that they are likely to build up higher levels of nitrosamines.) Nitrosamines can also build up in food in some circumstances. At one time, there was an intense debate about whether nitrosamines could be formed within the human body, but it is now generally accepted that this occurs.

Nitrates and babies

The best understood aspect of the health effects of nitrates is the risks which nitrates pose for very young infants.

High nitrite levels can be acutely dangerous to young babies because they can cause the oxidation of haemoglobin in the blood, bringing about the potentially fatal condition of methaemoglobinaemia, or cyanosis. This is also known as 'blue baby syndrome', because lack of oxygen in the blood turns the skin a bluish colour. Infants of less than three months are especially susceptible because, once the process has occurred, they cannot easily reverse it again.

There have been about 2,000 known cases of methaemoglobinaemia recorded worldwide since 1945. Only fourteen of these are recorded from Britain, the last known case being in 1972. Only one British case has been fatal, in the 1950s.

Other factors can also increase the risk of blue baby syndrome. These include bacterial contamination of water and lack of vitamin C. Methaemoglobinaemia is likely to be more acute in less developed countries, where pure water is not readily available and there is a lack of fresh food. It may have been increased by the promotion of powdered baby milk in these areas, because more babies will have been drinking dirty water, rather than mothers' breast milk.

The link between nitrates and methaemoglobinaemia led the World Health Organisation to suggest setting upper limits for nitrates in water as long ago as 1970. Since then, there have been various levels set by the United States health authorities, the EEC and national governments. The accepted safety level has fluctuated between 45 and 50 mg/litre.

Although it is well over fifteen years since a case of methaemoglobinaemia has been identified in Britain, the health authorities are increasingly worried by the potential risks, as nitrate levels continue to rise. The risks are considered serious enough for nursing mothers to be supplied with bottled water in parts of the country where nitrate levels in tap water are thought to pose a risk.

The Anglian Water Authority has two mobile bottling plants ready to supply bottled water to infants if nitrate levels should exceed 100 mg/litre. In the Yorkshire Dales, near Ripon, babies born in areas using borehole water had to be supplied with mains water from Harrogate until a new mains supply was installed in 1988.

However, it is not only full blown cyanosis which is a problem. Research in Israel suggests that methaemoglobin levels are significantly raised even in infants' drinking water containing nitrate levels permitted by the EEC. And studies in the USSR, looking at an older group of children who drank water with nitrates at only half the maximum EEC levels, again found elevated methaemoglobin levels in the blood, and tentatively linked this with slowing of the reflexes.

Nitrates and cancer

Although protection of infants is given as the main reason for setting limits on nitrates in drinking water, the possibility of a link between nitrates (or more directly nitrosamines) and the development of various forms of cancer is a much more contentious political issue.

Although much of the nitrate we eat is excreted again fairly quickly, some is reduced to nitrite, often by bacteria present in the body. Nitrite is formed wherever there are enough bacteria, including, on occasions, the mouth, stomach, intestine, bladder and vagina. For example, it has been estimated that about 5 per cent of dietary nitrate is converted to nitrite in the saliva. Some of this nitrite, in the form of nitrous acid, reacts with secondary and tertiary amines to form nitrosamines, and undergoes other reactions to form further N-nitroso compounds. Nitrosamines are formed in the saliva and, less commonly, in the stomach.

Nitrosamines are known to be powerful carcinogens (cancer forming agents). In 1981, Sir Richard Doll, the leader of the group which proved a link between tobacco and cancer, described N-nitroso compounds as being 'among the most powerful chemical carcinogens'. Nitrosamines have been found to be carcinogenic in 39 different laboratory animal species, including primates. Experiments have also shown that feeding nitrates and amines to experimental animals can lead to the formation of carcinogens identical to nitrosamines.

However, subsequent epidemiological research on human populations, and further experimental work, has failed to establish a link between nitrate intake and cancer in human populations, at least at the concentrations occurring in Britain.

Epidemiological research is, strictly speaking, the study of sickness in the population. Epidemiological studies are very detailed studies of large groups of people and of all the factors affecting them. For example, if there appears to be a high incidence of cancer among a particular group of people (e.g. workers in a factory, or people living in one particular area), and epidemiological study would look at all the factors which could be causing cancer and try to pinpoint the cause. When one particular carcinogen (cancer causing agent) is suspected, all the other possible factors also have to be taken into account, eg how much people smoked, what they ate, what other carcinogens they have come into contact with, and so on.

The main emphasis of epidemiological research into nitrates has been on investigating a possible link between nitrates and stomach cancer and nitrosamines. This is the second commonest type of cancer in the world, and one which is almost always fatal.

In England, initial studies took place in the Nottinghamshire towns of Worksop and Sutton-in-Ashfield, in East Anglia and in Yorkshire. Results were confusing. Some studies appeared to show an increase in stomach cancer in areas where nitrate exposure was higher, while others did not. No causal link was established either way.

One of the first detailed studies of nitrates and cancer anywhere in the world took place in the small Nottinghamshire mining town of Worksop. The population had an abnormally high level of stomach cancer, especially in women. Research showed that it also had an unusually high average nitrate concentration in the drinking water up to 1971. An initial comparison with nine other towns, which had lower nitrate levels in the water and lower incidences of stomach cancer, seemed to back up the theory of a link between the two.

However, a more detailed series of studies cast doubt on this link. A study of nearby Sutton-in-Ashfield found increased stomach cancer rates without the correspondingly high nitrate levels. A larger survey of ten Nottinghamshire towns concluded that, if social class structure was taken into account, along with the number of miners, the Worksop cancer levels were not particularly high after all.

Confusion and disagreement has continued to dog research into the possibility of a link between nitrates and cancer. And, as we shall see, social classes and other similar factors could themselves have an influence on the hazards from nitrates, if such hazards actually exist.

Studies in other countries added to the mass of data without answering any of the fundamental questions. Study areas included places in Denmark, Chile and Colombia. Some of these found that increased exposure to nitrates had occurred in areas with higher than average stomach cancer levels.

For example, a report from the World Health Organisation, *Guidelines for Drinking Water Quality*, published in 1984, reported that:

> In a review of gastric cancer in China, the Putian Prefecture of the Fujian Province was found to have the highest mortality from this disease (124–140 per 100,000 males). Data showed that in this area the levels of both nitrate and nitrite in drinking water and in vegetables were higher than in the low-risk areas.

However, other studies did not find any clear link. There has been a continuing decline of stomach cancer in Britain throughout the period when nitrate levels in water and food have been increasing.

Stronger evidence against the possibility of a nitrate-stomach cancer link came through research at the Imperial Cancer Research Fund (ICRF) Epidemiology and Clinical Trials Unit in Oxford, the results of which were published in the latter half of the 1980s.

One trial measured nitrate and nitrite levels in the saliva of 380 people, drawn from areas with high or low stomach cancer rates. Salivary nitrate levels were found to be significantly lower in areas with high stomach cancer rates. There is still no theory to explain this. The agrochemical industry (along with popular magazines like *The Economist*) hailed this as 'proof' that nitrates were safe. The authors of the paper were more cautious, writing that the results

> should not be taken to imply that nitrate-related N-nitroso compound carcinogenesis has no role in the development of gastric tumours, nor should they be taken to exclude the possibility that nitrates may play an important part in countries where salivary nitrate levels are much higher.

A second study at ICRF appears to have produced an even stronger case against there being a direct link. Sir Richard Doll and his team studied 1,327 people who had worked in an ICI fertiliser factory between 1946 and 1981. Again they measured salivary nitrate levels. The study showed two things. It proved that salivary nitrate levels are, indeed, related to nitrate exposure. The fertiliser workers averaged twice the salivary nitrate levels of nearby people who did not work at the factory. More importantly, the research team found no evidence of increases in either stomach cancer or in other forms of cancer, including cancer of the oesophagus and bladder. (This is important because oesophagal cancer has increased during the period of rising exposure to nitrates and a link between the two had been suggested.) Men who had been heavily exposed to nitrates for ten years did have a slightly increased risk of lung cancer.

These two are keynote studies. They suggest that, at the very least, the link between nitrates and cancer is nowhere near as clear cut as was once supposed. But, again, the experts themselves urge caution:

> the amount of nitrate to which people in England and Wales have been exposed is possibly insufficient to produce a detectable effect and much larger amounts may be needed.

So. . . . are nitrates safe?

So where does this leave us? Can we now disregard the issue of cancer altogether when arguing about the problems of nitrate pollution? Such complacency is premature. The evidence suggesting that nitrosamines are carcinogenic is unusually strong, making it worthwhile following up all available avenues of research. There are still a number of questions which have to be answered relating to a possible link between nitrates and stomach cancer.

These are, first, the reasons why stomach cancer is decreasing. In particular, whether an overall decline could be disguising a counter tendency for an increase in stomach cancers due to nitrates. Second, possible ways in which nitrosamines might be acting with other factors in promoting cancer. This could again obscure their role, and also mean that nitrates derived from different sources had differing effects. And third, the question of dose.

We will look at each of these in turn.

The decrease in stomach cancer

Stomach cancer has been declining in Western Europe and North America for the past half century. It is thought that this is due mainly to an increase in the quality of food production, storage and distribution methods, including especially:

- reduced mould contamination of food leading to decreased exposure to trichothecenes and carcinogenic mycotoxins (aflotoxin is one of these which has been in the news recently as a possible health risk from mouldy peanuts, etc);
- reduced intake of lipid peroxides in stored fats and oils;
- increased intake of phenolic antioxidants such as butylatedhydroxytoluene (BHT) which may be linked to a decline in deaths from stomach cancer (but may actually promote other types of cancer);
- increase in consumption of fresh vegetables (this is confusing, because vegetables are also the main source of nitrate, and is discussed in more detail below).

Thus there are many other factors affecting stomach cancer. Despite recent, justifiable, fears about food contamination, increased ability to store fresh foods has undoubtedly helped improve overall health in the rich countries.

Even if nitrates were promoting stomach cancer in some way, the increase could be obscured by a corresponding decrease caused by other factors. The decrease in stomach cancers observed in the population would, therefore, be less than would be the case without nitrates. This means that variation between and within various human social groups is likely to be complex because of all the other factors involved.

Links between nitrites and nitrosamines, and other forms of cancer, have only recently been explored in any depth.

The relationship of nitrosamines with other factors

The relationship between the (almost certainly) carcinogenic nitrosamines and other N-nitroso compounds, and other natural and artificial factors, is also important. Some of these might be able to 'activate' the nitrosamine's carcinogenic properties, while others might serve to protect against cancer.

In a recent survey of food adulteration, biochemist Eric Brunner of the London Food Commission wrote that:

> It is likely that N-nitroso compounds are important cancer agents in humans, while exposure to nitrates is not critical. Evidence suggests that nitrate is not the rate limiting . . . factor in human N-nitroso compound formation. Although nitrate is necessary for their synthesis, some other unknown factor may be the regulating one except perhaps at . . . extremely high nitrate levels . . .

If this is true, it makes interpreting effects even more difficult. First, it raises the possibility that N-nitroso compounds are only carcinogenic in certain forms, or when something else is present to catalyse the reaction. So, for example, nitrate in meat might be more dangerous than nitrates in vegetables; this is something we come back to below. Second, it means that, if nitrates are not the rate-limiting factor, then reducing them may not make much difference to overall effects.

There is also the possibility that any carcinogenic effects of N-nitroso compounds can be reduced by cancer-inhibiting agents. For example, as mentioned above, although fresh vegetables can have very high nitrate levels, they appear to inhibit stomach cancer. It is thought that this may be due to the anti-oxidant action of their vitamin C content. Any potentially harmful effects may be, on average, more than balanced out by the vegetables' ability to suppress stomach cancer. Good news for vegetarians. But, again, it makes detecting any overall pollution pattern uphill work.

Sociological and economic factors also come into the picture. If this premise were true, for example, it could help explain the increased incidence of stomach cancer amongst lower income groups. People on low incomes eat, on average, less fresh and frozen vegetables and greater amounts of processed meat like pies and sausages. Hence they have both less vitamin C and more additive nitrate and nitrite. (They also smoke more on average, further increasing their exposure to N-nitroso compounds.) Many questions remain to be answered.

The total dose of nitrates

Last, but by no means least, nitrate levels in the diet are still rising. Evidence from countries with considerably higher exposure to nitrates has suggested that there are links with stomach cancer, as described earlier. The ICRF researchers, whose work is central to the debate about nitrates and health, have continually stressed this point. Apart from workers in fertiliser factories (who may not fit other criteria of high stomach cancer levels), there have been few, if any, detailed studies of very high exposure to nitrates. So, despite all the hyperbole, we still don't really have a clear answer about whether nitrates cause cancer or not.

As always when there are high profits to be made, and investments to protect, scientific debate has been clouded with premature claims of safety, advertising campaigns and marketing managers seizing on any shred of evidence to promote the impression that nitrates are perfectly safe.

For example, the Fertiliser Manufacturers' Association published a pamphlet defending fertilisers in March 1986, *The Case for Fertilisers*, which bluntly stated that

> There is no evidence to connect nitrate intake with the incidence of cancer in the human population.

and also that

> Soil alone can release as much nitrate as is applied in fertilisers.

Such sectional interests inevitably stifle debate. Anyone genuinely worried about nitrates becomes loath to give ammunition to the agrochemical industry which could help produce whitewash. On the other hand, no one wants to waste time on 'problems' that do not exist.

Dr David Foreman, one of the co-authors of the papers about salivary nitrate experiments carried out by ICRF, summed up the caution still required at a conference organised by the Council for Environmental Conservation (now the Environment Council) in London in November 1986. He recommended avoiding high levels of exposure because of the methaemoglobinaemia risk and because

> we do not know about the long term effects. Generally I take the view that departures from biological normality should not be tolerated unless there is a good and positive reason for doing so.

Few people would disagree with the sentiments. However, the continuing uncertainty means that future decisions about nitrate levels in food and, to some extent, in water are likely to be even more controlled by political factors than is generally the case for environmental pollutants.

Certainly at the moment the British government appears bent on writing off the possibility of serious risks and assuming that nitrates and their by-products are safe. Let's hope they are right. In practice, because of the way in which nitrate limits are set to avoid methaemoglobinaemia, the debate about cancer has little effect on policy decisions about nitrate levels in water. However, the arguments could be crucial in determining the amounts of nitrates added to meat, or allowed to build up in vegetables.

Meanwhile, many specialists remain doubtful about the safety of nitrates. A report by the European Chemical Industry Ecology and Toxicolgy Centre, in Brussels, published in January 1988, still had doubts about the issues:

> Both N-nitrosamines and N-nitrosamides induce tumours in a wide variety of tissues in many animal species and at extremely low exposure levels. Although there is no direct evidence that N-nitroso compounds cause human cancer, it is unlikely that man is insensitive to their effects.

In 1987, the review of the International Agency for Research into Cancer's ninth meeting

> stressed that available biochemical and histopathological data give little reason to believe that humans are resistant to the carcinogenic action of NOC [N-nitroso-compounds].

And George Salt, Director of the School of Water Sciences at Cranfield Institute of Technology, summed up some of the remaining disquiet at a conference organised by CoEnCo (now the Environment Council) in November 1986. He pointed out that:

> It may be that some relatively unexplained incidence of illness will in time be discovered by the statisticians to be related to high nitrate intake . . . Nitrates have been rising steadily for some twenty to thirty years; we are about to expose a generation to a whole lifetime of high nitrates and it may be that only after forty or fifty years are we going to find the effects. So we simply cannot say with any degree of confidence what's going to happen, but I'm worried.

5: What nitrates do to the environment

As we have seen, the question about what damage nitrates in the diet can do to human health is so full of disagreement and uncertainties that it is still too early to draw precise conclusions. Although the same is true, to a certain extent, when we try to assess how much harm an excess of nitrates does to the environment, at least some of the issues here appear to be much more clear cut.

In this chapter, we look first at the range of environmental effects which occur when nitrates leach into freshwaters and give an overview of how much nitrate pollution actually occurs in British rivers and lakes. We then go on to look at the issues connected with nitrates in the seas around Britain, and, lastly, at their role in air pollution and acid rain.

Environmental impacts of nitrate pollution

The best known effect of fertiliser run-off, and of other forms of nitrate pollution, is eutrophication of water. Eutrophication is the extreme nutrient enrichment of water, which can result in water becoming clogged with algae, leading to oxygen starvation and the death of most other plants and animals.

In a natural river or lake, the amount of nitrogen and phosphorus present can often be the rate-limiting factors governing aquatic plant growth. Small additional amounts of nitrates and phosphates in freshwaters can, in theory, stimulate growth of aquatic flowering plants and result in lusher vegetation. However, overuse of fertilisers frequently means that overdosing of nutrients occurs.

When this happens, algae (and especially green algae) respond fastest to the sudden enrichment, and utilise the extra nutrients in a spurt of growth. They frequently undergo a dramatic population explosion in nutrient-enriched waters. If you see a stretch of freshwater become covered with a thick scum of green algae within a few days, or a 'bloom' of blue-green algae, there is a good chance that the water has been contaminated with excess nutrients from fertilisers, manure or human sewage.

The algal bloom has a number of effects. Sometimes the layer of algae becomes so thick that it blocks off light from submerged plants, preventing them carrying out photosynthesis. In extreme cases, the plants die. The remains of these plants, and of dead algae, are broken down by bacteria which themselves use up virtually all the oxygen present in the process. Most of the animals then suffocate as well, and the water becomes almost devoid of life.

In addition, blue-green algal blooms sometimes release toxins which kill fish and other animals. In a severely eutrophic freshwater, the only species which survive are a few water beetles and other insects, which breathe atmospheric oxygen by frequently swimming up to the water surface, and fly larvae known as rat tailed maggots, which breathe air from the surface through a long tube. Most of these animals feed on the remains of rotting plants or material which falls into the water and the freshwater ecosystem virtually dies.

For eutrophication to occur, both nitrogen and phosphorus have to be present. As nitrogen is often in considerable excess, attempts have been made to control eutrophication by controlling those phosphates which come from point sources such as industrial waste and sewage. However, it is much less easy to control phosphorus from fertilisers or manure.

So far so good. No one really disagrees with this. Where the arguments begin is on the question of the extent of eutrophication, and whether it is a serious environmental problem in Britain at the moment. For example the DoE report *Nitrates in Water* concluded that

> in general, the increased nitrate concentration presently found in some surface waters does not, by itself, adversely affect the ecological balance,

before going on to qualify this by admitting that

> in a few cases even small increases of nitrates have been shown to produce marked changes in the abundance of flowering plants, and the diversity and productivity of algae in lakes...Excessive plant growth is likely to be detrimental to aquatic life in general.

There can also be:

> . . . qualitative changes in algal communities, for example a switch from diatoms to blue green algae.

Other assessments have put greater emphasis on the ecological effects of eutrophication and have concluded that Britain faces a major problem with deteriorating water quality.

For example, in the Norfolk Broads, the reedswamp has declined dramatically in the last couple of decades, and nitrate levels have been pinpointed as the cause. The floating form of the reed *Phragmites* is particularly at risk, perhaps because the nitrates cause excessive top growth so that the reed literally falls over in the water. This has had wider environmental effects in the area, because the reedswamp formerly absorbed waves from passing boats and, with its demise, large scale bank erosion is occurring.

The exact scale of freshwater pollution, and the role of nitrates, remains in dispute. Nitrate is believed to be the rate limiting factor in the Lake District and in Loch Leven, in Scotland. The government's own Nature Conservancy Council has no doubts about the need for a reduction. Margaret Palmer, of the NCC, wrote in a letter to The Association of Agriculture at the end of 1988 that:

> In the opinion of the NCC the level of nitrates recommended by the EEC as a desirable maximum for drinking water is too high for the well being of wildlife. In the Norfolk Broads, concentrations of nitrate sometimes exceed 50 mg/litre but are often far lower than this. Whilst about 80 per cent of the excess phosphate and 20 per cent of the nitrate in our rivers and lakes originates from sewage works, about 80 per cent of the nitrate and 20 per cent of the phosphate comes from agricultural sources, including artificial fertilisers, slurry and silage. The number of incidents of water pollution resulting from spillages of slurry and silage is increasing.

Overall water quality

It is important to look at nitrate pollution in terms of overall water quality in Britain. Despite reassuring noises from government and industry, in many cases overall water quality is declining. In the Department of the Environment's 1985 survey of river water quality in England and Wales, for the first time in many years there was an increase in the lengths of polluted rivers, with nitrates and other agricultural effluents playing a major role in this deterioration. The North West Water Authority reported that

> over half the length of river stretches which deteriorated in quality (between 1980 and 1985) are on account of agricultural drainage both from diffuse and discrete sources. As an indication of the growth of the problem the number of successful prosecutions brought by the Authority for pollution offences resulting from less than adequate farm drainage has increased tenfold since 1980.

Many of these contamination problems also have an impact on smaller streams and drainage channels. These are often ignored in pollution statistics, but are nonetheless the habitat of countless numbers of aquatic animals and plants. The long-suffering Anglian Water Authority, writing in the same DoE report, pointed out that

> many small streams, particularly in Norfolk and Suffolk, were affected by agricultural run-off, causing a general deterioration in the underlying water quality.

There is now overwhelming agreement in Britain

that water quality is suffering as a result of nitrate pollution, and that this, in turn, is having a detrimental effect on aquatic life.

Nitrates in the sea

Although freshwater pollution remains an extremely serious issue in many areas, it may be less important overall than pollution of the sea around our coasts. Within the last couple of years, several marine mammals have suffered catastrophic declines. In 1988 alone, thousands of seals died of a mysterious virus, seabirds on offshore islands failed to breed properly, the last few families of dolphins raised no young and large algal blooms were found off the coasts of Scandinavia.

Marine life is suffering from many different problems: natural, climatic, pollution, bad management and overfishing. It is still difficult to distinguish the effect of any one pollutant. However, there is growing evidence that nitrates play a key role in the continuing decline of marine ecology around our coasts.

Nitrate pollution is already severe. Over 1.5 million tonnes of nitrogen are put into the North Sea every year from European countries, and two thirds of this comes from river inputs. Although the precise quantities derived from farming are not known, the total farming input can be judged to be enormous. The Ministry of Agriculture, Fisheries and Food's *Food Surveillance Paper number 20*, published in 1987, recorded that over 4,000 tonnes of nitrate entered British coastal waters every day. Important sources include fertilisers and untreated sewage sludge.

Sixty per cent of the anthropogenic nitrogen input to the North Sea comes from agricultural sources. This includes losses both from the land itself, and from factories making fertilisers (which are often near to the coast).

There is mounting evidence that these inputs are affecting water quality, especially in shallow coastal waters. In the sea, nitrogen is sometimes thought to be the rate-limiting factor on growth, rather than phosphorus as in the case of freshwaters, although our understanding of these processes is still very inadequate.

Whatever the precise causes, we know that eutrophication is increasing sharply in areas of the North Sea and the Baltic. As a result, over the last few years, the size and frequency of algal blooms have been increasing throughout the North Sea, and in shallow or coastal areas in the Irish Sea as well. Although there may be other factors involved, including other pollutants and climatic shifts, the role of nitrogen is increasingly being seen as a likely factor in these problems, and certainly can help to sustain and prolong blooms when they do occur. Unlike freshwaters, nitrates are often the limiting factor in the sea, where they are in smaller supply than phosphates.

A major algal bloom occurred off the coast of Scandinavia in May 1988. At one stage, a slick of algae 10–30 metres deep, and up to 10 kilometres wide, was snaking around the coast. Norwegian salmon and trout farmers faced losses totalling $200 million. In 1989, similar vast algal slicks affected the coast of northern Yugoslavia, around the Istrian peninsula, proving disastrous for the tourism industry of the area.

Eutrophication affects the marine environment in a number of ways. The most noticeable effects are amongst bottom living communities. In the Baltic, there has been a major decrease in bladder wrack around South Finland, and the upwelling of nutrient rich water is a probable cause. Increased nutrients increase populations of smaller algae, and of other plants and animals living on the seaweed. This in turn cuts down light intensity to the plants, and presses them to the bottom, thus preventing spore attachment. Increased numbers of small marine mammals, caused by the eutrophication, eat the bladder wrack and thus prevent regeneration.

These changes have knock-on effects on other forms of life as well. Plankton blooms are thought to have caused mortality in commercial mussel beds. There is increasing evidence that eutrophication is affecting fish populations in these areas as well, although this remains unproven. Occasional fish kills have been well documented, both amongst wild fish and caged fish and amongst lobsters in fish farms.

Oxygen levels in the bottom waters of the Baltic have decreased significantly since the 1950s, and the same effect may well be taking place in parts of the North Sea as well, leading to an overall reduction in productivity. In the Baltic, about 100,000 square kilometres suffer more or less permanent oxygen deficiency.

Effects on fish may also be occurring. Eutrophic waters can sometimes support large populations of bacteria, increasing the incidence of fish diseases. In the North Sea, algal blooms may also decrease the availability of fish spawning and nursery grounds as the algae settle on the sea bed. Some species produce mucus which is thick enough to clog up gills. Species identified as particularly at risk in the North Sea include sole, herring and, to a lesser extent, mackerel.

There is evidence that plankton blooms are increasing around the British coast as well, as was admitted by the government's own Nitrate Coordination Group in 1987. In the same year, the Oslo Commission reported that 'There has been an increase in nutrient levels in coastal waters around the UK in recent years. . .'.

We still don't know how serious these issues are going to be in the future, or the precise role that nitrates play in declining marine life. But there is a growing consensus that says evidence is now strong enough to merit action to reduce the pollution. Stanley Clinton Davis, the EC Environment Commissioner, specifically pointed out the risks excess nitrates pose to ocean life when he introduced new draft legislation curbing fertiliser use in December 1988:

> We have already seen the disastrous consequences of rising nutrient levels in the seas resulting in the blooms of algae in the North Sea, the Baltic and the Adriatic . . . This problem is on a Community scale and must be dealt with accordingly.

Similarly, the Draft Council Directive which Stanley Clinton Davis was introducing doesn't mince words about marine effects:

> Nitrate is particularly implicated in the eutrophication of marine waters and the subsequent occurrence of algal blooms. In recent years, such blooms have regularly occurred during the spring and summer months in the Baltic, North and Adriatic Seas. These blooms can cause great damage to the biota, particularly fisheries, and have a large negative effect on the touristic value of the affected areas.

We can expect the question of nitrate pollution of

marine waters to become an increasingly important factor in the debate about use of farmland in Europe.

Nitrogen and air pollution

Not all the waste nitrates from agriculture wash into rivers and the sea. A proportion volatilise, i.e. evaporate, into the atmosphere. Once there, they become an important constituent of the general air pollution found in industrialised countries, including acid rain.

One of the breakdown products of nitrogen found in animal manure is ammonium. In the last few years, there has been increasing evidence that ammonium volatilised from the large quantities of manure produced by factory farming units can be an important air pollutant. Although ammonia is not itself acidic, it can be broken down by soil bacteria into nitric acid, and leach into freshwater streams and rivers.

Ammonia is also volatilised from nitrate fertilisers, although less is known about this process. Hot, dry conditions, a high windspeed and alkaline soil are all known to increase emissions.

In addition, ammonium itself is a damaging pollutant to plants, including trees. One of the current hypotheses for forest dieback, or Waldsterben, is that trees are suffering from an overdose of nitrogen. Some of the forests which are now dying recorded record growth in the years before they started to decline. Dutch studies have linked pine needle loss and spread of fungal diseases in trees to poisoning by too much ammonia. The ammonia is hypothesised to increase losses of magnesium, potassium and calcium, and to build up nitrogen in the leaf, reducing frost resistance.

In the Netherlands, Dr N. van Breeman of the University of Wageningen found that up to three-quarters of the acids reaching soils in the Rouwkuilen nature reserve, which is surrounded by factory farms, came from ammonium sulphate. It's suggested that this is caused when airborne ammonia oxidises sulphur dioxide in the air to form ammonium sulphate on foliage. In Dutch forest soils at least half the acids appear to have come from nitric acid from ammonia.

Intensive livestock husbandry has been directly identified as a component factor in freshwater acidification and, especially, in tree dieback in countries such as the Netherlands. The Dutch government takes the risks so seriously that the further building or extensification of factory farm units has been banned until a solution is found.

In Europe as a whole, two thirds of ammonia emissions come from human and animal wastes, with a further contribution from nitrate fertilisers. 80–90 per cent of emissions are due to domestic livestock and fertilisers. There is growing evidence that serious ammonium pollution may be occurring in Britain as well. Ammonia emissions from British farms have increased by 55 per cent over the last 30 years, and currently stand at almost 400,000 tonnes a year. In areas with dense livestock populations, such as Cheshire, parts of Wales and central Scotland, ammonia emissions can exceed 5 tonnes per square kilometre per year. Emissions vary seasonally, and are largest in spring when spreading of wastes is common.

The increases from animal manure may in part be due to increases in livestock numbers, but probably owe more to the way in which manure is stored. In place of the traditional dung heap, most modern farmers store manure as a liquid slurry in open tanks, before spreading it onto fields. This increases the chances of volatilisation, as does the practice of keeping animals in confined spaces where manure is continually trampled.

Another important air pollutant caused in part by agricultural practice is nitrous oxide. Bacterial action on nitrogen fertilisers is thought to be a major source of nitrous oxides, and fertilised soils emit 2–10 times as much as unfertilised soils and pastures. Losses also occur from animal manures.

Like ammonia, levels of nitrous oxides are rising, currently at a rate of 0.2–0.3 per cent per year, and by 2030 or so levels could be half as much again as in pre-industrial periods. Nitrous oxide is an important air pollutant, and plays a major role in acid rain formation in some areas.

6: Kicking the habit: cutting down nitrates in our diet

I have tried to be fairly scrupulous in listing the ambiguities about the effects of nitrates on health and environment. Nonetheless, like most responsible governments and the European Community, I believe that it is important to reduce the amounts of nitrates both in our own diet and in the environment as a whole, and especially in freshwaters. The next three chapters give some suggestions about how this can be achieved.

Chapter 6 starts at the personal level, by looking at how to cut back nitrates in our own diet, from water, meat and vegetables. Chapter 7 looks at ways of reducing nitrate losses from water, once pollution has occurred. Chapter 8 looks at both the possibilities of changing conventional farming systems and at more radical methods for cutting back nitrate losses, by a conversion to organic agriculture.

As we have seen earlier, nitrates are always present in vegetables in small amounts, whatever the growing method used. They will also almost always be present in significant quantities in water in industrialised countries. No-one should expect, or even attempt, to reach a completely 'nitrate free' diet. However, for people wanting to reduce their exposure the following steps should be able to help.

Water

The government estimates that 1.3 million British people currently receive some of their drinking water with nitrate levels exceeding the maximum laid down by the EEC. Many others are close to the limit, and Andrew Lees of Friends of the Earth puts the total figure at nearer 4 million. High nitrate levels are commonest in the east of England, but also occur in parts of the south and Midlands. Private wells in agricultural land may also have especially high nitrate levels. The worst times of year for nitrate pollution are likely to be winter months, particularly after heavy rain, when surges can be expected.

If you are going to be raising an infant on bottled milk diluted with water, and are in an area known to have high nitrate levels in water, it is worth checking with the local health authority to see if you are eligible for bottled water to reduce the baby's nitrate intake.

Although filters are sold to purify tap water, they are unlikely to reduce nitrate levels very much, whatever their other benefits. Recent research by the Consumers' Association, reported in *Which?* magazine, found that only one of the filters they tested removed nitrates from water, and this was only effective when the filter was fairly new (i.e. only for the first 40 litres). Some manufacturers sell separate nitrate removing cartridges for their filters, which should be more effective, although these were not tested by the Consumers' Association.

It is certainly *not* worth boiling water. This doesn't

get rid of nitrates and, in fact, a study at the University of Iowa suggests that nitrate levels are concentrated in boiled water, increasing the risk. The only real option is to replace tap water with purchased mineral water, or tap water drawn from a region where nitrate levels are known to be lower. However, this will prove expensive if mineral water is also used for tea, coffee etc.

It is also worth noting that mineral water is itself sometimes polluted with other substances. For example, some European mineral water was recently measured with radiation levels seventeen times the World Health Organisation's recommended limits, and high levels of bacterial contamination are by no means uncommon. Because mineral water is not chlorinated, bacteria levels can rise steeply if it is stored in shops or in the larder for some time before being drunk. The Consumers' Association found a high incidence of bacterial contamination of mineral water, in tests conducted in early 1989.

Meat

As we have seen, food additive nitrites in meat and dairy produce may be amongst the more hazardous sources of nitrates (although it must be stressed again that this is still only one theory amongst several). It is certainly a source which could be avoided with a fairly small change in official policy. In the medium term, it is important to persuade the government to replace the additive nitrites E249 and E250 with safer alternatives. For example, it is thought that by using ascorbic acid or scorbic acid the use of nitrates could be considerably reduced; other options might cut it out altogether.

For the present, options are limited unless you are a vegetarian. Organic meat exists, but supplies are still very limited. There are also slightly larger supplies of various 'additive free' meat which, whilst not fully organic, is produced from animals which are not fed hormones or antibiotics, and where meat doesn't have nitrates added. Recent changes in European Community policy on agriculture, including funding for extensification, may increase the amount of organic meat available.

Vegetables

People worried about the (sometimes very high) nitrate levels in vegetables are faced with a dilemma. Vegetables are also believed to be a major source of protection against the possible side-effects of high nitrate intake. In general, advice must be to *eat a good proportion of fresh vegetables in the diet*. However, protection against nitrate side effects will be increased even more if vegetables liable to contain high nitrate levels are avoided. These include:

- vegetables known to concentrate nitrates, such as *spinach* and *lettuce*, which have been grown out of season or under glass;
- vegetables known to concentrate nitrates when grown in conditions of heavy artificial fertiliser input.

Note that whilst organic vegetables avoid the second criteria, and should thus be bought when possible, any nitrate-storing vegetables grown under glass are likely to build up high nitrate levels, whether or not the system is organic.

Knowledge about vegetables likely to build up high nitrate levels is still incomplete. They include *spinach*, *lettuce*, *celery*, *fennel*, *beetroot*, *turnip*, *radish* and *cornsalad*. A more complete list appears in Chapter 3. People worried about nitrates in vegetables could reduce the risks by not eating vegetables out of season.

To sum up. . . .

If you live in an area with high nitrate levels in the water, there don't appear to be many ways of avoiding this unless you get a good water filter and constantly change it, or by expensive (and dubious) mineral water. The only ways of avoiding nitrates in meat are either going vegetarian or buying organic, or additive free, meat. You're better off eating vegetables, but can reduce the risks of eating high nitrate levels with them if you only buy the 'high risk' vegetables when they are in season.

7: Kicking nitrates out of water

At the moment, the problem of high nitrate levels in water is tackled, if at all, by removing a proportion of the offending nitrates from drinking water before it reaches our taps. Current plans are apparently for an increase in this clean-up, to the minimum level that the European Community will allow Britain to get away with.

So far, Water Authorities have adopted three strategies for reducing nitrate levels in drinking water:

(1) substituting water from other sources;
(2) blending high and low nitrate sources to produce an acceptably low average;
(3) treating water to reduce nitrate concentrations.

Each of these is examined briefly in the following paragraphs.

Utilising cleaner water sources

Substitution of water has taken place in areas where groundwater sources exceed the World Health Organisation (and now also the EC) limits. Substitution has been confined to providing nursing mothers with bottled water to make up milk for their infants, if required. Until recently, this has been assumed to be the cheapest method of cleaning water and has probably helped prevent the spread of methaemoglobinaemia. However, it does not provide any safeguards for adults (or for children once they stop being tiny babies) and costs rise very steeply as soon as the number of nitrate-polluted sources increase.

Blending water also takes place in Britain, and is a fairly cheap alternative when high and low nitrate sources exist close to one another. Unfortunately, high nitrate levels tend to be regional and blending becomes much more expensive if water has to be transported large distances.

Denitrification

This leaves us with water treatment, or more precisely denitrification, as the method for tackling the nitrates problem. Although some denitrification occurs when water is simply left to stand, especially if there are plenty of plants growing in the water to take up nitrates, the problem is now too acute for this to provide a sufficient cure. A number of methods have been devised to speed up the natural process of denitrification. The two best known are biological denitrification and ion exchange.

Biological denitrification takes advantage of naturally-occurring micro-organisms which can reduce nitrate to nitrogen under anoxic (low oxygen) conditions. A carbon source, such as methanol, usually has to be added before the reaction will take place on a large scale.

A full-scale plant handling 2.3 billion litres of water a day is being evaluated in eastern England, using a fluidised sand bed for the removal of nitrate and methanol as the carbon source. Some concern has been voiced over the toxic effects of methanol, and the whole process is currently being monitored. Alternative carbon sources such as acetic acid are also being investigated.

Ion exchange* has been developed mainly for use with groundwater supplies, because in these sources the low organic content makes biological denitrification unsuitable. In addition, groundwaters do not usually require the other treatments given to surface waters, into which biological denitrification can fairly easily be incorporated. This makes it a more expensive process for groundwater.

The resins available for ion exchange have a poor selectivity for nitrate. Costs can vary considerably if other ions, especially sulphate, are also present in the water being treated. Large quantities of regenerant are required in relation to the nitrate removed, and disposal of spent regenerant is a major part of the total cost involved. Reverse osmosis has been investigated as a method of reducing the amount of regenerant required.

The ion-exchange process is quite resource intensive, and generates large amounts of waste material which then have to be disposed of. Two or three plants are being built in the east of England, mainly for

---

*_Ion exchange methods_ utilise the ability of an ionised surface in water (or another ionising solvent) to exchange its own ions for others of the same sign, supplied by suitable electrolytes. It is used commercially with _ion exchange resins_, which are polymers with a particularly high exchange capacity. Ion exchange methods are used in a wide variety of applications, including analytical processes, catalysis and water treated. In the latter use, if a mixed bed of anions (negatively charged atoms) and cations (positively charged atoms) exchange resins are used, electrolytes can be removed from a water solution, and very pure water obtained.

groundwater, although one station will denitrify river water.

Costs of reducing nitrates

Potential costs for reducing nitrates to the level demanded by the EC were reviewed by the Nitrate Coordination Group. They estimated that to meet the EC's 50 mg nitrate/litre maximum would result in £200 million capital costs, along with running costs rising to £10 million a year. Since the NCG reported, the number of people receiving water with unacceptable levels of nitrates has risen from 1 million to 1.3 million, meaning that these costs are already a considerable underestimate.

Should nitrate be removed from water at all?

Until recently, it has been assumed that cleaning up nitrates from water was the responsibility of those controlling Britain's water supply. Although this was in clear contravention of the Polluter Pays Principle, it was justified by arguing that this was the cheapest method of control, as compared to controls on nitrate fertilisers which might reduce crop yields. This view was taken by the Royal Commission on Environmental Pollution in 1979 (the first major British report which covered the nitrates issue) and again by the Nitrate Coordination Group in 1986, who wrote that 'it is not easy to apply the Polluter Pays Principle effectively and fairly to nitrates'.

Although it has never been explicitly stated as such in official reports, this decision amounted to another hidden subsidy for Britain's farmers.

The NCG based their 1986 recommendation on an analysis carried out by the consultants Lawrence Gould and Associates. When the NCG report, *Nitrates in Water*, appeared, there was a collective groan from all those who had been lobbying for a more a more equitable and environmentally benign system. However, within just a few weeks of the report's appearance, and despite all the agonising which undoubtedly went into the text, the Lawrence Gould proposals appeared to have been dropped. Civil servants in MAFF, at least unofficially, were agreeing that economic criteria were not the only important factors and that control at the source of pollution was also essential.

This is probably the most fundamental change to have taken place in official attitudes towards nitrate pollution over the past ten years. We can now look sensibly at controlling nitrate pollution where it should be controlled; by changing the wasteful way in which we practice intensive farming in much of Europe.

8: Kicking the habit in agriculture

Once it has been decided that nitrate levels in water need to be reduced, and that polluters should be responsible for this reduction, then there are two basic strategies that can be followed.

First, remedial steps can be taken within a conventional farming system, to reduce the level of nitrate leaching and other nitrate losses. This is the option which most farmers will opt for in the short term, if stricter controls on nitrate losses are introduced. Such 'add-on' management strategies are sometimes lumped into a rather vaguely defined concept known as low input farming.

Second, a more radical approach can be taken, and the farm switched to an organic system of agriculture. This is an integrated ecological approach to agriculture which, if practised correctly, includes reduction of nitrate leaching amongst its benefits. Organic farming practices will be examined in more detail below.

Reducing nitrate pollution from conventional farming

A number of simple guidelines have been drawn up, to reduce nitrate losses on chemical farms. Some of these are minor management strategies, others involve more fundamental changes in the way that farming is arranged. Few would include input from people and institutions beyond the farm gate. Despite the millions of words written and spoken about nitrate pollution, precise knowledge about what these changes would mean in terms of reducing nitrate leaching is still not clearly understood. The commonest options are listed below:

(1) Working out nitrate requirements much more accurately on a field by field basis, rather than applying an excess as a matter of course. A number of factors have to be taken into account, including crop uptake, existing nutrient levels, type of fertiliser, alternative sources, etc.

(2) Applying nitrate fertiliser as late as possible in the spring, and not at all in the winter, to avoid the highest rainfall, when nitrate is heavily leached from soil.

(3) Splitting application of spring fertiliser, to reduce the excess at any one time, or using slow release fertilisers if time or cost dictates that there is only one application.

(4) Not applying nitrate fertilisers to autumn crops unless this is absolutely essential for growth.

(5) Planting winter cereal as early as possible in the autumn.

(6) Maintaining high levels of carbon in the soil, to reduce the risk of excess nitrates. One method is to plough straw waste in during the autumn, rather than burning off the stubble. The straw 'mops up' excess nitrogen and releases it slowly during the next growing season.

(7) Planting a cover crop after harvesting, and not ploughing this in until temperatures are low

enough to reduce the decomposition rates of crop and root residues. This avoids the release of a surge of nitrates.

(8) Converting arable land to permanent pasture in areas where water systems are especially prone to nitrate pollution. This can include developing permanent buffer zones of grass and/or trees along sensitive lakes and streams to intercept run-off from adjacent land.

(9) Using soil conservation techniques and mulches to reduce run-off from adjacent land.

(10) Cutting down the number of livestock in sensitive areas and applying the 'polluter pays' principle more strictly against livestock farmers who cause pollution from burst slurry tanks, etc.

(11) Treating animal waste in sewage plants in exceptional circumstances.

(12) Controlling the total amount of nitrate fertiliser applied.

Until recently, almost all farmers would have regarded much of this as heresy. Some still do today. Indeed, when the government's own Agricultural Development and Advisory Service (ADAS) suggested that use of winter fertiliser application on bare fields be discontinued, there was a storm of protest. However, this has now generally been agreed, even by the Fertiliser Manufacturers' Association.

Despite the objections from the farming establishment, it should not be thought that these ideas are the ravings of green utopians. Almost all come from a paper produced by the staunchly pro-business and conservative Organisation for Economic Co-operation and Development (OECD), and published as long ago as 1973, called the *Impact of Fertilisers and Agricultural Waste Products on the Quality of Waters*. Some, but not all, were endorsed fourteen years later by the Nitrate Coordination Group.

Despite the complaints of farmers who oppose any attempts to reduce their output, controls on nitrate pollution are almost inevitable. They will probably be linked with more general strategies to reduce crop surpluses. Discussion now centres on which methods to go for and how to implement and enforce any measures taken.

Current thinking within Britain seems to be centring along a number of options which we shall examine in turn: changes in agricultural practices to reduce nitrate losses; a protection policy for certain sensitive areas (this is now an integral part of the 1988 Water Privatisation Bill); limitation on nitrates available to farmers, through quotas or taxation; and encouragement of alternative activities on farms.

Changes in agricultural practice

The Nitrate Coordination Group is the latest and (in Britain) the most heavyweight group to make recommendations about ways of changing farming methods to reduce nitrate leaching. The NCG proposed a number of steps including:

- no application of nitrate fertiliser in the autumn, from mid-September to mid-February;
- planting autumn-sown crops instead of spring-sown crops;
- planting winter cereal as early as possible in the autumn;
- maintenance of crop cover during autumn and winter;
- careful assessment of fertiliser needs on an individual field basis;

- leaving grassland unploughed.

Although people lobbying for proper nitrate controls were glad to see some of these proposals included, there was scepticism about the relevance of the recommendations in practice. There was also disappointment at some of the measures that the NCG felt had 'little potential for reducing nitrate leaching significantly'. These include, amongst others, slow release fertilisers, direct drilling, straw incorporation (as opposed to stubble burning), alternative crops, and a fertiliser tax. The NCG were also very sceptical about the potential role of organic farming in reducing nitrate losses.

However, other organisations remain more optimistic about the potential of at least some of these steps. There are also signs that MAFF has changed its mind considerably about several recommendations since the Nitrate Coordination Group report was published.

Protection Policy

The idea of operating a 'protection policy' around sensitive areas, such as boreholes, was assessed by Lawrence Gould Associates for the DOE in 1985. It was referred to in the NCG report, without any very clear recommendations being made except that the government should 'explore the practicality of protection policies'. In fact, the government has had the legislative powers to set exclusion zones ever since the 1974 Control of Pollution Act. These powers have been retained in the 1988 Water Bill.

The issue is difficult from an environmental point of view. Whilst there are undoubtedly areas which require special protection, an emphasis on 'special sites' risks a policy which allows business as usual everywhere else. With our land use policy already heavily biased towards protection of key areas rather than an overall approach, protection policies for nitrate control are likely to receive a lot of attention. However, they may not fit so easily into the overall changes which are required of agriculture over the next few decades.

The 1988 Water Privatisation Bill has strengthened the chances of water protection zones, and of some funding being available to utilise organic farming methods in these situations (see below). The scope and practicality of exclusion zones has yet to be tested, but they look likely to go ahead anyway. Draft EC legislation, released late in 1988, will require member states to designate 'all zones vulnerable to water pollution from nitrogen compounds', within which applications of livestock manure and 'chemical fertilisers' will be restricted.

Nitrate limitation

Various methods of limiting nitrates have been discussed by both the Nitrate Coordination Group and the National Farmers Union. The NCG were sceptical about nitrate taxes (i.e. controlling nitrates by making them more expensive, perhaps on a sliding scale). However, this option receives support from a fair number of farmers, if the letters columns of the farming press are anything to judge by.

The NFU issued a paper which discussed various options for nitrate control in 1987. It looked at a number of options for deciding how nitrates should be 'allocated', including on past use, per unit area and a standard ration per hectare for particular crops and livestock. The discussion paper tended to come down on the side of past use; good for the already successful large farmers but a serious disadvantage for new farmers or farms who wish to convert land to another crop to meet changing demands.

Nitrate limitation is quite a likely alternative for the near future. It should be carefully assessed from outside the agricultural power base, otherwise it could well be used to support the rich and powerful farmers to the detriment of their smaller neighbours.

Changes in land use

This is another tricky option. Environmentalists, country lovers and others would almost certainly welcome changes in farm use, but probably not in the way that the government is planning at present. Changing farm use involves both temporary or permanent changes to the use of land whilst maintaining it as farmland, and changing land use altogether.

Recent European Community plans have looked at various possibilities for 'extensification' to reduce food surpluses. Concrete proposals currently centre on paying Community farmers to 'set aside' a proportion of their land for non crop uses. In practice this has meant mainly for short-term fallow. The British government is also encouraging tree planting on farms as an extra option. The potential remains for farmers to switch their whole farm to a less intensive system, such as organic farming. Although this possibility has frequently been discussed, it has run into serious problems with vested interests within the agribusiness world.

Proposals published in 1989, for extensification aimed at beef cattle, do give more hope that less intensive systems could be adopted in these particular cases. However, beef cattle provide relatively few problems with respect to nitrate pollution anyway.

Changes in land use are being touted around as the solution to a whole range of problems at the moment, ranging from grain mountains to rural unemployment. It is unlikely that they will solve any of these on their own. In the case of nitrate pollution, current proposals are heavily biased towards changing land use in marginal areas, i.e. precisely those places where heavy use of nitrate fertilisers has never really been adopted anyway. They are, therefore, unlikely to make any major impact on nitrate pollution in Britain as a whole.

Less acceptable options for change of land-use include taking land out of production altogether and putting it down to new building, golf courses, leisure parks, sports facilities etc. These solutions, whilst reducing nitrate losses, would have many other side effects which could well be more serious. If they are adopted, it will not be because of nitrate pollution, but for other social and economic reasons, and are thus beyond the scope of the current book.

Organic farming: the radical alternative

We can expect a good many of the 'containment' methods described above to be introduced into British agriculture over the next few years. Indeed, a few are already in motion. However, if nitrate pollution is as chronic a part of the modern farming scene as it now appears, these first aid measures will not be enough to regain the clean rivers which most people would like to see flowing again.

An understanding of the real scale of the problem has brought demands for a more radical change in farming methods. A growing number of people want to see a move away from an agriculture dominated by chemical inputs, and towards a more ecological approach. Active development of this alternative has already been under way for some time in Britain, carried out by the thousand or so farmers who have partly or completely changed over to an organic system of agriculture.

It is important to define exactly what organic farming entails and how it differs from conventional farming.

Modern intensive farming methods rely on heavy use of artificial chemicals to mould the soil environment into a condition that will allow maximum growth of crops and animals. This involves, in particular, adding liquid fertilisers to provide an abundance of nutrients, and adding pesticides to kill any animals or other creatures which feed on the crop, or weeds which compete for space. Although chemical farming has resulted in high yields, it brings a number of damaging side effects, including breakdown of soil structure, adulteration of food and pollution from pesticides. Nitrate pollution is one result of this effect.

Organic farming, on the other hand, relies predominantly on an understanding of ecology rather than just a knowledge of chemistry. It tries to strike a balance between changing the environment to fit the crops and adapting farming methods to fit a particular environment. This results in slightly lower yields than in chemical-based farming, but maintains a far more robust and stable growing system, and avoids many of the problems of environmental destruction and food contamination which accompany chemical farming.

Over the last few years, organic farming has lost much of its traditional folksy image, although this picture is perpetuated by the chemical farming lobby as much as possible. Despite the idea that many people have about organic farming being a step back towards how farming used to be, modern organic methods are still being developed, and it remains a very new system. Important constituents of organic agriculture include:

(1) Introduction of crop rotation to reduce both build-up of crop specific pests and the development of nutrient imbalance where certain crop species require large amounts of a particular nutrient.

(2) The use of plant and animal composts and green manures as sources of soil nutrients. Composts are made by mixing waste material and allowing it to decompose into a state that is easily used by growing plants. Green manures are crops sown purely to add to the nutrient content of soil, especially through the use of legumes which can fix nitrogen from the air.

(3) Control of pests by biological methods, barriers, intercropping and other tactics which do not include use of hazardous pesticides.

(4) A comprehensive set of standards for livestock husbandry, including more humane treatment and a ban on growth stimulating hormones.

Although yields would be lower in an organic system, there would not necessarily be very substantial changes in total food output. For example, recent calculations by Nic Lampkin, of the Department of Agricultural Economics at Aberystwyth University, suggest that wheat yield in a 100 per cent organic Britain would be 70 per cent of current levels. In practice, organic farming will be introduced gradually and, as it is still being developed, yields should rise even further as more is learned about techniques.

Does organic farming reduce nitrate pollution?

Organic farming is certainly better for the environment as a whole. Recently, it has been put under scrutiny to see if it is also better from the point of view of nitrate losses.

In an organic system, run in the way laid down by the Soil Association Symbol Standards, use of soluble nitrate fertilisers is banned, and application of animal manures carefully controlled. The main source of nitrates within the organic system is a grass-clover ley, introduced into the rotation once every four years. The clover, being a legume, fixes nitrogen from the atmosphere.

The elimination of soluble nitrate fertilisers, and the inclusion within the rotation of years when the soil is not cultivated, both combine to produce relatively low leaching rates in an organic system.

Most research into nitrate losses on organic farms has, to date, concentrated on specific parts of the rotation. However, scientists at Elm Farm Research Centre, an independent organic research farm, have recently assembled this data into a picture of average nitrate losses over time. Their results appear to be confirmed by a study over the whole period of rotation, undertaken at one organic farm. Overall nitrate leaching rates are apparently much reduced. Experiments at Elm Farm Research Centre, the government's own North Wyke Research Station, and elsewhere, have shown that annual average leaching losses of as low as 20 kg/ha are achievable in a correctly-managed organic system. Losses can rise to 100–150 kg/hectare at the time of ploughing in the ley; however, in other years losses from a grass/clover ley are as low as 5kg/ha.

In addition, manures are usually used very carefully on an organic farm, so that losses are minimised. Autumn applications are avoided as a matter of course, and in general manures and slurries are applied to grass rather than arable crops.

These results show the importance of involving practising organic farmers in research work of this kind. Simulations undertaken by scientists who do not understand the practice of organic farming have sometimes been simplistic or misleading. For example, the use of manures on a working organic farm is very different from that tested recently at Rothmanstead Research Station, where manures were applied in large quantities to bare, arable land and high leaching rates recorded. Yet these results were widely quoted as 'proving' that organic agriculture also resulted in high nitrate leaching. More careful research shows that the reverse is, in fact, the case.

The role of organic farming in nitrate limitation

It is difficult to know how far the government will back organic farming as a policy option for reducing nitrate pollution. The assessment made by the Nitrate Coordination Group was fairly pessimistic about its role, but subsequent discussions between representatives of organic growing organisations and civil servants have been much more positive.

During 1989, several statements from both the government and the Labour Party have appeared to support grants to help farmers over the crucial conversion period, when they change their land over to an organic system. The Labour Party has pledged support for such a conversion grant under a future Labour administration. The EC extensification programme may also help organic farmers more than it has in the past. However, these helpful noises have not, as yet, resulted in any positive help for Britain's organic farmers, or for chemical farmers wishing to convert. Some of these policy issues are examined in more detail in the following chapter.

9: The future of nitrate control

I started this book by arguing that the nitrates problem had been around for a long time, but had suddenly leapt into prominence and thus become an important political issue for environmental groups (and a fitting subject for a book). This sudden interest has not been spirited out of a vacuum. Some of the pressures were briefly discussed in the first chapter; the new regulations coming out of the EEC, changes in agriculture throughout Europe and water privatisation in Britain. Each of these topical policy issues has a direct and important bearing on what will happen with regard to nitrates in the future. This last chapter makes an attempt to summarise a few of the more important issues.

The role of the EEC

The control of water pollution in Britain has been defined by five successive Acts of Parliament, and by English Common Law. Until recently, only two of those Acts were of real importance: the 1973 Water Act and, to a lesser extent, the 1974 Control of Pollution Act. Both these are vague in their requirements. The Water Act set a number of obligations, including that the Secretary of State is obliged to secure a national policy for water including restoration and maintenance of the wholesomeness of freshwaters. What exactly 'wholesomeness' means is not at all clearly defined. It has resulted in control of water pollution being, to a large extent, governed by the kind of old boys' network of voluntary agreements much beloved by both civil servants and industry in Britain.

Now all this has changed. As the European Community introduces a steadily growing number of Directives on water pollution and other environmental issues, the control of Britain's water purity is increasingly being taken out of our own government's hands. The 1988 Water Privatisation Bill, which will from now on be the major piece of legislation governing water pollution, has been written with EEC regulations in mind.

Some of the relevant Directives are listed in Table 5 overleaf. By far the most important, from the point of view of water quality and nitrates, is the Drinking Water Directive COM (75) 394. This Directive sets 62 water quality standards, along with many guidelines for water quality monitoring. Three types of standards are set: Guide Levels (GLs) which are voluntary; Maximum Admissible Concentration (MAC); and Minimum Required Concentration (MRC), both of which are legally binding and have to be met by all Member states.

The Directive had enormous implications for the UK because, for the first time, it has set mandatory limits for a variety of pollutants. This means that we now have something resembling a definition of water qual-

Table 5: General areas of interest with respect to water quality controlled by European Community Directives

General area	*Scope of Directive*
Drinking water	Directives on: drinking water; surface water used for drinking; sampling surface water used for drinking; and groundwater.
Fish	Directive on water standards for freshwater fish.
Detergents	Five inter-related Directives.
Dangerous substances	A general Directive, plus Directives on: mercury (2 Directives dealing with the chloralkali industry and other sources); cadmium; and titanium oxide.
Pesticides, etc.	Directives on: lindane; DDT; carbon tetrachloride; and pentachlorphenol; and (proposed) aldrin, dieldrin and endrin.
Marine pollution	Directives on: shellfish waters; bathing water; and oil pollution at sea.
Information	Directive on Exchange of Information – Water, with amendment, to establish exchange of information on the quality of water in the Community.

Source: *The Potential for developing organic agriculture as a mainstream policy option and reducing surpluses and protecting the environment*, edited by Lawrence Woodward and Christopher Stopes, Elm Farm Research Centre, 1988.

ity in Britain, which the government and industry cannot fudge over by, for example, altering or bending voluntary upper limits for pollutants. Thus, for nitrates, the EC set an upper limit, or MAC, of 50 mg/litre for any drinking water supply.

The British government's response

Not that this has stopped the British government from attempting to introduce a 'fudge factor' or two into the proceedings. The government has chosen to regard some of the MAC values as absolute, while others have been defined as an average over a period of time. Thus, for nitrates, a three month average is taken, during which levels on individual days can (and sometimes do) exceed those defined as the maximum by the EEC, although this is clearly against the spirit of the Directive.

In addition, the Directive sometimes allows exemptions (called 'derogations') for pollutants to exceed their MACs, although only in certain very carefully defined circumstances. This is usually either in times of emergency, or where 'adequate treatment is not possible', generally taken to be where some pollutants occur in naturally high concentrations. However, the UK has applied for derogations for over 200 water supply zones in England and Wales where pollution is anything but natural. 57 of these are for nitrates.

Worst of all, the British government arbitrarily 'redefined' the upper limit for nitrates from 50 mg/litre to 80 mg/litre, higher than for any other country which has set limits at all. They argued that this was justified because nitrates are not a 'toxic' substance, but this interpretation was later abandoned under pressure from the European Court.

Like many Community Directives, whilst the back-up documentation and discussion make the meaning clear, the actual wording of the Directive is ambiguous in places, and leaves a lot of room for discussion. (This happens precisely because Member States want some leeway for manoeuvering.) Nonetheless, there is a general consensus that Britain overstepped the mark on this occasion. The British government was taken to the European Court of Justice over its non-compliance with the Drinking Water Directive, by Friends of the Earth and by a number of individual complainants, and has subsequently had to make some more concessions.

In December 1988, the British government was put under further pressure by the publication of another proposed Directive, which will add substantially to the controls already imposed on nitrates. The draft Directive sets out a framework for action, but puts the onus on individual member states to come up with specific programmes for 'vulnerable zones' where the problems of nitrate pollution are most severe. Amongst the controls envisaged are:

- clear limits on the amount of animal manure farmers could apply to land;
- appropriate limits on chemical fertilisers, including control on waterlogged, flooded and frozen ground;
- controls on dumping of municipal sewage in vulnerable zones from a population of 5,000 or more to ensure that it does not exceed a concentration of 10 mg/litre.

Thus the European Community will apparently continue to increase pressure on the British government to take the nitrates issue more seriously than it has done until now.

Nitrate sensitive areas

To date, the main response of the British government to pressure from the EEC to reduce nitrate pollution has been the planned introduction of nitrate sensitive areas, or NSAs, where nitrate use will be far more tightly controlled. In a discussion document published in May 1989, the Ministry of Agriculture laid down plans for a series of experimental NSAs, where a range of agricultural measures to reduce nitrate pollution could be tested out. NSAs are to be in 'areas where nitrate concentrations in water sources exceed or are at risk of exceeding, the limit of 50mg/l . . .', but will first be tried out in a series of pilot areas representing a range of conditions likely to be found in the rest of the country.

The NSA scheme is seen as a three stage process: an advisory campaign by the Agricultural and Development Advisory Service (ADAS); a series of voluntary measures; and, if necessary, compulsory measures, with compensations for farmers involved.

Some of the measures approved for use in NSAs look set to be contentious. They include a number of commonsense steps such as: avoiding autumn fertiliser or manure applications; planting winter cover crops; avoidance of grassland ploughing; and ensuring that livestock densities are low enough that sufficient land is available for manure spreading without overdosing.

However, they also suggest that leguminous forage crops should be avoided, and that grassland should be unfertilised, and without clover. Both these measures effectively ban the practice of organic agriculture on NSAs, despite clear evidence that it reduces nitrate leaching. Legumes are an essential part of any organic rotation, as they provide the fertility-building phase of the cycle. And keeping grassland free from clover, a natural plant, would entail the use of weedkillers banned in an organic system. Thus if the NSAs are adopted as planned, organic farming to Soil Association Symbol Standards will be effectively banned.

Elm Farm Research Centre staff write, in a joint response to the paper from several organic growing organisations, that:

> The opportunity for resolving the problem through a rational de-intensification of agriculture using organic farming systems is not given scope in the document and we consider this to be regrettable.

Privatisation of water

In 1985, the Conservative government announced that they intended to privatise the Water Authorities, as the next stage in their obsession with creating a free market in all areas of British life. The proposals caused an unprecedented storm of protest. The trade union NALGO organised a mass-based campaign to oppose the moves, drawing in other trade unions, environmentalists, health groups and organisations as diverse as the Womens' Institute and the Soil Association.

There were a number of reasons why water has become a much more controversial privatisation issue than, say, British Telecom or the National Bus Company. First, there are no real precedents for a privatised water supply. Even in the USA, the global nerve centre of capitalism, the water is controlled by the state. There is a certain amount of built-in resistance to throwing such a huge undertaking into private hands. Second, and probably more important, widespread and well-articulated fears quickly emerged to the effect that selling off water was a sure-fire formula for selling off much of our control over water quality itself. (There is also a deep-seated gut feeling against water privatisation; how can you possibly privatise something which falls out of the sky on its own accord?)

Initially, the problem seemed to rest with the way in which the newly privatised water companies would be controlled. As investors would, in practice, mainly be large transnational companies, there seemed to be no way to prevent a major polluter from buying into the water company and thereby gaining a lot of control over how pollution was (or was not) controlled in 'their' rivers and streams. Like selling the banks to the robbers as someone put it. So great was the weight of public opposition, that the government dropped the issue until after the next General Election took place in 1987.

However, with another large majority, and Mrs Thatcher continuing her radical political restructuring at full tilt, it was inevitable that water privatisation would again reach the political agenda in Britain. This time, the Conservatives had thought the issue through more carefully and proposed a centralised Rivers Protection Board to undertake the policing and implementation of careful pollution and environmental legislation. The immediate environmental issues seemed to be at least partly answered and many conservation organisations, whatever their members' individual feelings, temporarily withdrew from the debate, leaving trade unions increasingly isolated. A Bill to introduce the privatisation of water was introduced in December 1988.

Whatever the problems or merits of privatisation from a political or social point of view, the selling off of the Water Authorities will have one spin-off which would be very relevant to the nitrates issue.

We have already seen how the Water Authorities are subsidising the farming community by covering the cost of removing nitrates from drinking water sources. This is in clear contravention of the 'polluter pays' principle which is supposed to apply in all areas of British life. The anomaly has only survived because the Water Authorities have been publicly-owned bodies, controlled by successive governments intent on improving agricultural productivity and prepared to pay a high price to achieve this.

In privately-owned water companies this cosy relationship will no longer operate. Rapidly burgeoning legislation regarding water pollution means that the owners of Britain's water supply system are going to have to pay considerably more to clean up their 'product' than they may have bargained for. When there is such a clear source of the pollution as is the case with nitrates, the private companies are probably going to insist that the major part of these costs are borne by the polluters, in this case by the farmers of southern and eastern England.

Response from the farming community

It is easy to satirise the farmers' viewpoint into one of solid opposition to anything that reduces their crop yields, and hence their own profits from the land. Easy but incorrect. The debate about nitrate pollution and the inevitability of limitations has been a constant feature of the farming press, and farming organisations, in recent years.

In general, it would be fair to say that the farming press has been resigned to nitrate limitations, rather than enthusiastic to change agricultural policies. An article on nitrate in the *Farmers Weekly* in September 1988 is not atypical in leading off with

> Nitrates in UK drinking water supplies are rising above the EC health safety limit. There is no known scientific basis for the EC figure of 50 mg/litre and the clean-up will cost more than £200 million. It could mean a ban on arable systems in some areas. (We) reveal a prodigious and costly political folly which threatens us all. Brussels insists, so who will pay the bill?

However, there is also growing acknowledgement that, even if the exact cut off point is in dispute, nitrate levels are rising so fast that current debates will be superceded by more urgent problems if action is not taken now. David Naish, the National Farmers Union deputy president, spelled out the reality facing farmers in an interview in *NFU Insight* in December 1987

> UK water nitrate levels are rising and could, if no action is taken, reach 150–200 ppm in certain parts of the country, which is too high a level to be acceptable to anyone.

The *Farmers Weekly* leader represents the nub of the farmers' worry; whether or not the clean-up is needed, will they get lumbered with the costs?

In conventional logic, as outlined earlier in this book, the farmers who cause the pollution should, indeed, pay the bill, according to the 'polluter pays' principle. Other industries which unknowingly cause pollution face the costs of cleaning up, and there have been ample warnings about nitrate pollution for years. Yet the farmers do have a point when they claim that successive governments have devised entire agricultural policies around the concept of maximising yield, and that the farmers should not carry the entire burden of repairing any damage. Over the past few years, there has been a large change in attitude amongst many farmers, angry about swings in government policy and feeling sold out by the agrochemicals industry. The desire for a new emphasis on quality food production, with minimum environmental disruption, is growing.

It would also be wrong to give the impression that farmers are only looking at tinkering with existing practices. The need for, and desirability of, limitations and a more general de-intensification have been appearing constantly in the pages of the farming press for some time. Northumberland farmer Aidan Harrison put the issues starkly at a meeting on nitrate limitation at the Royal Agricultural Society centre at Stoneleigh, Warwickshire, as long ago as 1986:

> The growth of Europe's farm surpluses is comparable to an accelerating car. But the measures the Eurocrats are using to try and slow it down are like letting the air out of the tyres, removing the sparking plugs or disabling the driver. It would be more sensible to ease up on the accelerator.

The meeting preferred a nitrate limitation policy to one of setting aside agricultural land.

Changes to farming in Europe

To a large extent, plans to control nitrate pollution within the context of current agricultural practice are already out of date. Due largely to mismanagement of the Common Agricultural Policy, Europe has been producing too much of certain key agricultural products for many years, leading to the infamous 'grain mountains' and 'wine lakes'. In a world where most people are undernourished, the primary agricultural problem currently facing Europe is how to cut down production without destroying the agricultural community in large areas. Farming is going to have to change. As we have seen that nitrate pollution comes about largely because of intensification of farming, it would seem logical that any changes in the future should include steps to reduce nitrate leaching.

Unfortunately, the fact that something appears logical is no guarantee that it will occur. The current debate about the future of European farming involves powerful interests which are fighting hard to turn agriculture in a direction that will provide them with maximum profits, rather than in ways which might provide wider benefits.

European policy makers are starting to grapple with the next stage of the extensification programme. Extensification is used here as a catch-all term for ways to reduce the amount of food we produce in Europe. Although government and industry do not put it this way, extensification is in effect a reversing of the policy of intensification, practised for the last forty years in Britain.

Out of the mass of discussion papers, research documents and proposed Directives considering extensification, three main options have emerged, although there are overlaps between all of these:

(1) Removal of land permanently from agriculture by using it for building houses or factory units, sports facilities such as golf courses, or making areas into woodland, conifer plantations, national parks or nature reserves.

(2) Removal of land temporarily from agriculture, also known as 'set aside'. Paying farmers to take a proportion of their land temporarily out of production, and not use it for other crops in surplus.

(3) De-intensification of farming, which would reduce output more generally, whilst maintaining roughly the same amount of land under cultivation.

The first option is already happening to some extent, with fresh grants available for planting both broad-leaved woodlands and conifer plantations, and by housing developments on new green field sites. All these options are controversial, but are beyond the scope of the current book. In any case, taking land permanently out of cultivation is no solution to the immediate problems of surpluses. The areas being converted are not large enough to make a real dent in the grain mountains, and much of this land is poor quality and therefore not used for products which are in surplus.

In general, support for the other two options divides fairly neatly between environmentalists on one side and bureaucrats and industrialists on the other, with the farmers who will actually have to implement them left somewhat undecided in the middle. From an environmental point of view, a less intensive farming system (and ideally a completely organic farming system) is the best option for many reasons, including the likely effect of reducing nitrate leaching and nitrate pollution. From a civil servant's point of view, de-intensification is more difficult to categorise neatly into packets suitable for grant aid, and MAFF officials also think it would be more difficult to police. Industrialists have no reason to do anything but oppose moves that would reduce the need for their products.

At the moment, it is difficult to judge exactly what will happen. However, some de-intensification seems likely, if only because grants are likely to dry up for some of the more wasteful practices. Marginal land areas that have been pulled into production in the last few years are not unlikely to revert to rough ground again (or disappear under a thick mass of conifers). The role of de-intensification on prime agricultural land, where nitrates are a far more serious problem, is still a complete unknown.

10: Charters for the future of nitrate control

Two important charters have recently been published from within the environmental movement, which relate to nitrates and nitrate control. The *Friends of the Earth Charter for the Water Environment* was launched in September 1988 as a precursor to a major campaign in water quality in Britain. Some of the parts of the charter most relevant to the current book are reproduced below. In 1989, the *Soil Association Charter for Nitrates in the Farm Environment* was launched to draw attention to the steps needed in reducing nitrate pollution from agriculture. It is reproduced in its entirity, and is launched to coincide with the publication of this book.

The Friends of the Earth Charter for the Water Environment

1. Better safeguarding of the public's health

a. The application of pollution prevention measures, including the designation of water protection zones, to provide more effective controls over land use practices and discharges which affect water quality.

b. Making all drinking water, including supplies from private sources, comply with the legal limits set in the EC Drinking Water Directive and, for parameters with no specified Maximum Admissible Concentration, the World Health Organisation's guideline values.

c. Treating drinking water to remove all pollutants which pose any threat to the public's health.

d. Comprehensive and statistically sound monitoring of the quality of drinking water by both water undertakers and district councils.

e. Public disclosure of detailed information about drinking water quality.

2. Promoting of wildlife and landscape conservation

a. The classification of river and other water bodies according to scientifically sound, ecological principles.

b. The recognition of wildlife conservation as a recognised use of water in the definition of river quality objectives.

c. Ensuring that all waters comply with the EC Directives on dangerous substances in water and the EC Directive on the quality of water required by freshwater fish.

d. Full appraisal of all developments affecting water through the application of the EC Environmental Impact Directive and public disclosure of the details of all environmental impact and cost/benefit appraisals relating to such developments. . .

3. Pollution prevention and prosecution for all pollution offences . . .

c. Environmental standards for substances in water determined by an appraisal in public of the scientific evidence of the levels necessary to safeguard the environment.

d. The full implementation of the 1974 Control of Pollution Act.

e. The removal of Crown Exemption to make all discharges to water subject to the 1974 Control of Pollution Act, EEC law and the Paris Convention.

f. The application of the 1974 Control of Pollution Act to all groundwaters through their designation as 'specified underground water'. . . .

h. The full enforcement of statute law and the full use of civil law to safeguard rivers and other aquatic habitats . . .

j. Statistically sound monitoring of the water chemistry, sediment chemistry and aquaitc flora and fauna (species/community structure and bioaccumulation) of all waters which are affected by discharges of pollutants from point and diffuse sources.

k. Sufficient scientific staff and other resources for effective water pollution control. . . .

(A full text of the Water Charter is available from Friends of the Earth, 26–28 Underwood Street, London N1 7JQ; telephone 01–490–1555.)

The Soil Association Charter for Nitrates in the Farm Environment

In the last few years, organic farming methods have been carefully analtysed with respect to their impact on nitrate leaching. The available evidence now suggests that a correctly managed organic farm will suffer substantially less nitrate leaching than a chemical farm. The Soil Association, and other organic organisations, are calling on the British government to introduce organic farming as a serious option in nitrate sensitive areas and in other places where nitrate limitation is considered to be important. The following charter sets out these ideas in more detail.

The potential for organic farming to reduce nitrate pollution and allow food production in nitrate sensitive areas

Current plans for reducing nitrate leaching are inadequate

Current plans for reducing nitrate leaching, including the introduction of nitrate sensitive areas, will not solve the problem of nitrate contamination of surface and groundwaters. They are seriously flawed in a number of ways:

- By concentrating only on areas currently in crisis, they fail to stop the general increases in nitrate leaching which are being experienced in many other agricultural areas.
- Many of the limitation methods involve taking the land out of production altogether, thus seriously disrupting the lifestyles of people currently farming the land, and probably losing the potential to farm on the land forever.

- Other limitation methods are unproven, or are unlikely to result in sufficient reduction in nitrate leaching.

Organic farming systems

The Soil Association is interested in the potential of using organic agricultural methods as a more sustainable and equitable method of reducing nitrate pollution.

- Organic farming systems seek to avoid, as far as possible, the use of outside inputs. In general, nutrient supply, weeds, pests and disease control are achieved by rotational practices including legumes, recycling of manures and vegetable residues, variety selection and the creation of a diverse ecology within and around cropped land.
- If outside inputs are required, then the least environmentally disruptive – at soil, plant and human level – are used. These inputs are typically, but not invariably, relatively insoluble for nutrients, and naturally occurring in the case of pesticides.
- The main objective for organic farmers is to establish an agronomic balance between the regenerative phase of the rotation, involving fertility-building and the breaking of weed and disease cycles, and the exploitative phase, involving cash cropping.
- Unlike conventional arable systems, where the soil is cultivated in most years, the rotational practices involved in an organic farm reduce the number of years of cultivation through the inclusion of leys within mixed ley/arable rotations. In addition, soluble fertiliser inputs are not used. These character-

istics can result in relatively low leaching rates for nitrates.

Organic farming systems and nitrate leaching

There is now growing evidence that use of an organic system results in substantial reductions in average nitrate leaching over the course of the rotation.

- Monitoring work conducted over the course of a whole rotation on an organic farm suggests that mean nitrate leaching levels of approximately 20kg/hectare are achievable. Within nitrate sensitive areas, such leaching levels would be consistent with a reduction in the levels of nitrate entering ground and surface waters.
- Organic farming systems have a number of other environmental benefits, including reduced soil erosion, the elimination of artificial pesticides and the creation of habitats suitable for wildlife on the farm. Organic agriculture also produces food which is currently in demand, and in short supply.

Organic farming systems as a way of reducing nitrate leaching

Organic farming systems therefore offer a middle way forward in the nitrates debate. They maintain land in agriculture, and produce an increasingly popular product, whilst reducing nitrate leaching to an acceptable levels. They also fulfill a number of other environmental criteria, and help reduce national surpluses without an additional, expensive set aside policy.

The Soil Association therefore proposes that:

- Organic farming is included as a major option in

nitrate sensitive areas, and in other areas where nitrate limitation is already of critical importance.

- Farmers in other areas are encouraged and helped to convert to an organic system by the re-allocation of some of the agricultural support grants towards the conversion and maintenance of an organic system.
- Financial support is provided for further research into various aspects of organic agriculture, including more work on the potential for using organic agriculture as a means of nitrate limitation.
- Organic methods are included amongst low-input techniques in educational programmes aimed at farmers with a view to reducing nitrate levels.

11: Finding out more

As far as I know, this is the first complete book for the general public on the nitrates question. On the other hand, there is a vast, and growing, literature on the issue in the scientific press, government publications and the transactions of learned societies. Some of the best and/or most comprehensive sources are listed below.

Three official or semi-official reports put the nitrates issue on the map. Although I would not agree with everything they said by any means, they are all worth reading for anyone interested in looking into the issue in detail. They are:

The Royal Commission on Environmental Pollution, Seventh Report: Agriculture and Pollution, published in 1979 and with the Commission chaired by Sir Hans Kornberg. This study looked at the impact of fertilisers used in modern farming systems on water and food. Some of the questions it raised still remain unanswered.

The Nitrogen Cycle of the United Kingdom; A Study Group Report of the Royal Society, published by the Society in 1983. Good background, though overly cautious in its assessments.

Nitrate in Water: A Report by the Nitrate Coordination Group for the Department of the Environment, published by HMSO in 1986 (although not actually released until 1987). A long delayed report, which again gives good background. As discussed above, the conclusions already appear to be out of date in terms of thinking within the DoE and MAFF.

In addition, there are a number of reports and conference proceedings produced by voluntary organisations and environmental groups looking into the issue.

Nitrates in Food and Water, by Nigel Dudley, London Food Commission, 1986. This report forms the basis of much of the present book, and contains a great deal more material as well. It should be consulted for a detailed reference list. However it is now slightly out of date, and has been updated by Eric Brunner as part of a new book from the LFC:

Food Adulteration and How to Beat It, The London Food Commission, published by Unwin Hyman in 1988. Also contains much useful information on additives, pesticides, irradiation and other issues.

Danger! Additives at Work, by Melanie Miller, also published by the London Food Commission, contains additional information on nitrates in meat.

Countdown on Nitrates, the proceedings of a conference organised by the Council for Environmental Conservation (now known as The Environment Council) in November 1986, which included papers on many of the aspects of nitrates discussed in this book. It is edited by Amanda Nobbs and Fiona Jeffries Harris and is available from The Environment Council at 80 York Way, London N1 9AG.

The Nitrates Story: no end in sight, by Vogtman and Biederman, gives detailed information about nitrates in vegetables and the potential role of organic farming in reducing nitrate levels. It is published by Elm Farm Research Centre.

Because this book is a short introduction rather than a scientific treatise, I have only included key references as footnotes. Most of the books above have long reference lists for anyone wanting to follow up certain points in more detail.

Over the past few years, a number of environmental and organic growing groups have studied the nitrate issue in detail. These include:

Elm Farm Research Centre: Hamstead Marshall, Newbury, Berkshire RG15 0HR, telephone 0488–58298. Have carried out practical and policy research into the role of organic farming in reducing nitrate leaching from farming systems.

Friends of the Earth: 26–28 Underwood Street, London N1, telephone 01–490–1555. Produce many detailed briefs on the issue.

London Food Commission: 88 Old Street, London EC1V 9AR, telephone 01–253–9513. Published one of the first reports on nitrates in food and water.

Soil Association: 86 Colston Street, Bristol, BS1 5BB, telephone 0272–290 661. Have carried out a great deal of policy research and lobbying relating to nitrate levels.

Glossary of technical terms

Amines: Organic compounds derived from ammonia.

Aquifer: Layer of rock or soil capable of holding and/or transmitting large amounts of water, used here in relation to groundwater (*qv*).

Biosphere: The region of the earth's surface and atmosphere in which life is found.

Blue Baby Syndrome: common name for methaemoglobinaemia (*qv*).

Carcinogen: a substance which is capable of stimulating the development of one or more types of cancer (e.g. nicotine).

Cyanosis: another name for methaemoglobinaemia (*qv*).

Epidemiology: strictly the science of epidemics. In this case, it means studying the effects of varying levels of nitrates in large populations, to determine whether they show any difference in rate of cancer, etc., when compared with population groups with different intakes of nitrates. Also involves taking all other possible factors into consideration to make sure that observed effects are really caused by the factor being studied.

Eutrophication: excessive enrichment of freshwater with nitrates and phosphates, leading to a population explosion of algae and consequent oxygen depletion when bacteria break down the algae once it has died.

Groundwater: water found in underground aquifers (*qv*), traditionally very pure and supplying 30 per cent of Britain's drinking water.

Legumes: plants of the pea and bean family, which have nitrogen-fixing bacteria in their root nodules, enabling them to collect nitrogen from the soil.

Methaemoglobinaemia: a condition of infants which occurs when nitrites combine with haemoglobin, forming methaemoglobin, which has a much lower oxygen binding capacity and can result in death through lack of oxygen.

Nitrates: Salts of nitric acid. Sodium and potassium nitrates are used as a source of nitrogen fertiliser in conventional agriculture.

Nitrites: Salts of nitrous acid. Nitrates can be converted to nitrites in the body. Nitrites are also sometimes present in small amounts in drinking water and are used as meat preservatives.

Nitrosamines: Substances arising from the interaction of nitrous acids with secondary or tertiary amines (*qv*). Capable of being produced from nitrates via nitrites in the human body. Most are highly carcinogenic.

Organic farming: an integrated ecological approach

to agriculture which, amongst other things, cuts out the use of artificial fertilisers.

Surface waters: Any freshwater found on the planet's surface, i.e. streams, rivers, lakes, ponds, etc.

Volatilise: to evaporate rapidly.